The Electrician's Trade

DeMYSTiFieD®

DeMYSTiFieD® Series

Accounting Demystified
Advanced Calculus Demystified
Advanced Physics Demystified
Advanced Statistics Demystified
Algebra Demystified
Alternative Energy Demystified
Anatomy Demystified
asp.net 2.0 Demystified
Astronomy Demystified
Audio Demystified
Biology Demystified
Biotechnology Demystified
Business Calculus Demystified
Business Math Demystified
Business Statistics Demystified
C++ Demystified
Calculus Demystified
Chemistry Demystified
Circuit Analysis Demystified
College Algebra Demystified
Corporate Finance Demystified
Databases Demystified
Data Structures Demystified
Differential Equations Demystified
Digital Electronics Demystified
Earth Science Demystified
Electricity Demystified
Electronics Demystified
Engineering Statistics Demystified
Environmental Science Demystified
Everyday Math Demystified
Fertility Demystified
Financial Planning Demystified
Forensics Demystified
French Demystified
Genetics Demystified
Geometry Demystified
German Demystified
Home Networking Demystified
Investing Demystified
Italian Demystified
Java Demystified
JavaScript Demystified
Lean Six Sigma Demystified
Linear Algebra Demystified
Macroeconomics Demystified
Management Accounting Demystified
Math Proofs Demystified
Math Word Problems Demystified
MATLAB® Demystified
Medical Billing and Coding Demystified
Medical Terminology Demystified
Meteorology Demystified
Microbiology Demystified
Microeconomics Demystified
Nanotechnology Demystified
Nurse Management Demystified
OOP Demystified
Options Demystified
Organic Chemistry Demystified
Personal Computing Demystified
Pharmacology Demystified
Physics Demystified
Physiology Demystified
Pre-Algebra Demystified
Precalculus Demystified
Probability Demystified
Project Management Demystified
Psychology Demystified
Quality Management Demystified
Quantum Mechanics Demystified
Real Estate Math Demystified
Relativity Demystified
Robotics Demystified
Sales Management Demystified
Signals and Systems Demystified
Six Sigma Demystified
Spanish Demystified
sql Demystified
Statics and Dynamics Demystified
Statistics Demystified
Technical Analysis Demystified
Technical Math Demystified
Trigonometry Demystified
uml Demystified
Visual Basic 2005 Demystified
Visual C# 2005 Demystified
xml Demystified

The Electrician's Trade DeMYSTiFieD®

David Herres

New York Chicago San Francisco Athens London Madrid
Mexico City Milan New Delhi Singapore Sydney Toronto

Cataloging-in-Publication Data is on file with the Library of Congress.

McGraw-Hill Education books are available at special quantity discounts to use as premiums and sales promotions, or for use in corporate training programs. To contact a representative please visit the Contact Us page at www.mhprofessional.com.

The Electrician's Trade DeMYSTiFieD®, Second Edition

2 3 4 5 6 7 8 9 0 DOC/DOC 1 2 0 9 8 7 6 5 4

ISBN 978-0-07-181887-2
MHID 0-07-181887-1

Sponsoring Editor
Judy Bass

Acquisitions Coordinator
Amy Stonebraker

Editing Supervisor
David E. Fogarty

Project Manager
Vastavikta Sharma,
Cenveo® Publisher Services

Copy Editor
James Madru

Proofreader
Yamini Chadha,
Cenveo Publisher Services

Production Supervisor
Pamela A. Pelton

Composition
Cenveo Publisher Services

Art Director, Cover
Jeff Weeks

Cover Illustration
Lance Lekander

My close friend, Judi Howcroft, renowned nature photographer and friend of the earth, took the photographs that appear in this book, and it is with deep appreciation that I acknowledge her daily inspiration and insight that made this project a reality.

About the Author

David Herres is the owner and operator of a residential and commercial construction company. He obtained a Journeyman Electrician's License in 1975, and has certificates in welding and wetland delineation, along with experience with elevators. Beginning in 2001, Mr. Herres has focused primarily on electrical work, upgrading his license to Master status. The author of *2011 National Electrical Code Chapter-by-Chapter* and *Troubleshooting and Repairing Commercial Electrical Equipment* (also from McGraw-Hill Education) has written over 100 articles on electrical and telecom topics. Mr. Herres resides in Clarksville, New Hampshire.

Contents

Introduction

Early humans observed in the world around them electrical phenomena—lightning, electric eels, static charges—but many centuries passed before ways were found to make use of them. In the last 150 years we have progressed from the first attempts at filament lighting to a situation where, as seen from outer space, the dark side of the earth sparkles with energy conveyed by mammoth interconnected power grids.

To the passing alien, lighting is the most visible part of the equation. Additionally, there are powerful electric motors that lift us to our workplaces in high-rise buildings, move materials, saw wood, and perform countless industrial operations

Electricity produces heat, dries grain, conveys information, and empowers our lives.

We are living in an extraordinary wired environment. But all this technology cannot sustain itself untended. The worldwide power grid and connected equipment are like an enormously friendly and useful workhorse that has to be fed and cared for if it is going to perform as expected. This is what this book is all about.

How to Use This Book

Create a plan! You'll optimize your experience if you go beyond just an impressionistic reading of this book and come up with a program. Approaches will differ depending on the individual reader's needs and goals. My objective has been to provide two primary benefits. First, you'll want to comply with your local electricians' licensing agenda. Rather than giving an exhaustive listing of all requirements everywhere (which would take up half this book), I have focused on a limited number of examples so as to provide an overview of the licensing process.

Then I show how to do an online search to find contact information, fees, work experience required, and exam description for the state(s) in which you plan to work.

A significant part of the book is devoted to the *National Electrical Code (NEC)*, just as major portions of all electricians' licensing exams focus on the Code. Accordingly, we discuss the Code's organization, structure, and ways to access exam answers on an open-book basis.

The other big part of this book talks about how to become a successful working electrician. Here we temporarily turn away from the *NEC* and discuss other tools of the trade, physical and intellectual. There are chapters that look at both these areas. Some of the material is strictly practical and hands-on, and some is of a more theoretical nature. These always work together. What you do with tools on the job is directly connected to knowledge that you possess, regardless of whether it comes from books or actual work experience.

We'll work together to create a relevant learning experience. To this end, multiple-choice and open-ended questions will be found in each chapter with answers in the Answers section at the end of the book. These short exams should be taken on an open-book basis so that they resemble licensing exams. This means that if you are not completely certain of the correct answer, feel free to go back over the chapter to find what you need to know. After you have completed the test, go to the Answers section and grade yourself. In line with many state licensing exams, the passing grade is 70 percent. If you do not pass at first, do not despair. Just go back over the chapter and then retake the test. You may want to keep at it until you score 100 percent.

Also at the end of the book, there is a much longer multiple-choice and true-false NEC Practice Exam based on the *National Electrical Code*. Here also the passing grade is 70 percent. It is an open-book exam. You should have your copy of the Code before you so that you can consult it as needed.

A couple more introductory notes:

To assimilate this book, score well in licensing exams, and become a successful electrician, advanced mathematical knowledge is not essential. Calculus and analytical geometry can be left to the engineers. You need to be adept at number crunching with the aid of a handheld calculator, and you must be able to manipulate simple high school–level algebraic equations, transposing to solve for various unknowns. A little elementary trigonometry is helpful in understanding some concepts that have to do with alternating-current (ac) waveforms, but this knowledge is optional.

If you need to review elementary mathematical material, I highly recommend the works of my McGraw-Hill colleague Stan Gibilisco. I must say that his writing is incredibly clear and lucid and very much in keeping with the subheading for all books in the Demystified series, which is "Hard stuff made easy." A listing of all his works may be viewed and the books may be purchased by typing his name into the Amazon.com search bar. Some especially useful volumes are

- *Everyday Math Demystified*, Second Edition
- *Algebra Know-It-All*

As for electrical knowledge, it is assumed that readers of this book understand fundamentals that are familiar to most people—the difference between alternating current (ac) and direct current (dc), how two wires are necessary to complete a circuit, and that sort of thing. State licensing exams generally include a section on electrical theory, and to score well on it, you need a certain amount of more advanced knowledge. I'll have a lot to say on that subject in the pages that follow. In this connection, I recommend two additional McGraw-Hill books by Stan Gibilisco.

- *Electricity Demystified*, Second Edition
- *Electronics Demystified*, Second Edition

Concerning the *NEC*, the current edition (or whatever previous edition is enacted into law in your jurisdiction) should be purchased immediately (price discounted at Amazon.com) because it is frequently referenced in this book and needed for your open-book licensing exam.

A previous McGraw-Hill book written by me, *2011 National Electrical Code Chapter-by-Chapter*, will be useful to interpret and learn to navigate the Code so that correct answers can be located quickly in preparing for a licensing exam or when job-site questions arise.

Bon voyage…

David Herres

chapter 1

Licensing for Electricians

The purpose of electricians' licensing is to protect the public from ignorant or unscrupulous individuals who would do substandard work. And certainly programs of this sort are effective. The concerns that come to mind immediately are the twin demons of electric shock and fire, but there are other issues as well. An electrical installation should be efficient and durable and have a good appearance, and it is in the interest of the community that its buildings are not only free of hazards but also of high quality in other ways.

CHAPTER OBJECTIVES

In this chapter, you will

- Examine and compare the electrical licensing requirements of two typical states.
- Learn the differences between journeyman and master (or supervising) electricians.
- See some common deficiencies found by electrical inspectors.

Typical Licensing Programs

One of the objectives of this book is to provide the information you need to obtain an electrician's license. Right off we have to recognize that various countries, states, municipalities, and other jurisdictions do not have identical requirements. Many of them are similar, but some are far more complex and extensive than others. Certain states (among them some of the larger, including New York and Illinois) have no electricians' licensing at all, ceding the whole matter over to municipalities or counties. New York City has a robust electricians' licensing program. Two levels of electrician's licensing are offered: master and special. Unlike most jurisdictions, the exam is not open-book, and it covers the *New York City Administrative Code* rather than the *National Electrical Code* (NEC), so it is unusual in these respects.

TIP ***Be sure to find out what materials you can bring into an open-book electricians' licensing exam. That and other pertinent information will be found on the Electricians' Board website for your jurisdiction or by calling the Board.***

We'll look at electricians' licensing requirements for two representative states, New Hampshire and Oregon. Both these jurisdictions have extensive user-friendly websites with educational information and documentation concerning licensing requirements.

New Hampshire Electricians' Licensing

New Hampshire electricians' licensing is administered by the Department of Safety. Individuals are required to be licensed to perform electrical installations within the state. *Electrical* is defined narrowly as pertaining to lighting and power only. Accordingly, one could work on signaling and communications circuits in New Hampshire without a license. For example, satellite dish installation is not regulated because power for such systems is provided by merely plugging the modem into a wall receptacle. Grounding of the system does not require an electrician's license because that is not considered a power or lighting issue. New Hampshire is in the process of developing a fire alarm technicians' licensing program, but as of 2013, the certification is voluntary.

As in many jurisdictions, home owners may do electrical installations and repairs on their own primary residences without an electrician's license, but (to take an extreme example) an unlicensed developer could not wire an apartment building or subdivision regardless of ownership. Carpenters and similar tradespeople without an electrician's license can legally do limited amounts of work on power and lighting systems. The work must be incidental and encountered within the course of practicing their principal trade. For example, in remodeling a room, a carpenter might rewire a single switch or receptacle. However, this individual cannot do work that involves calculations, such as extending a branch circuit to add a new outlet. Trunk slammers whose advertising lists various handyperson activities and includes "light electrical work" are definitely in violation. This is so because they are soliciting electrical work rather than encountering it in the course of their principal trade.

New Hampshire recognizes and licenses three levels of electricians. The usual entry-level permit is the *apprentice card*. The initial fee is $30. No prior technical knowledge or experience is required. The applicant must list the current employer and most recent former electrical employer. It is necessary to divulge any felony conviction(s) and whether a previous right of apprenticeship has been suspended, revoked, or sanctioned in any jurisdiction. A high school diploma is required and must be listed in the application. This requirement is waived if the applicant is a registered student in a youth apprenticeship program for high school students. A photo must be included.

The apprentice card allows the individual to install and repair electrical wiring and equipment in New Hampshire only under the supervision of a journeyman or master electrician. The supervising electrician must be on the job site any time the apprentice is on the job. Moreover, there must be one journeyman or master electrician for each apprentice present at all times. Electrical inspectors visiting the site make sure that the one-to-one ratio is not exceeded. In addition to these job-site requirements, each apprentice must be employed by a master electrician whose name is to be provided in the initial application.

Apprentices must satisfactorily complete a minimum of 150 hours of electrical schooling in an approved training course during each 12-month period in which the apprentice identification card is valid. On receipt of certificate, transcript, or official letter of enrollment in or on completion of electrical schooling, the State of New Hampshire will continue to renew the permit every year until a journeyman electrician's license is acquired.

Still Struggling

It is not necessary to know every requirement for every license. Just research what is needed for a license in your state and meet the requirements.

It is in the interest of the apprentice to upgrade to journeyman status as soon as it is possible to meet the requirements, including passing the exam. A journeyman electrician is valuable to an employer because rather than requiring one-on-one supervision, a journeyman can actually supervise an apprentice, helping to maintain the mandatory ratio. However, a journeyman can do electrical installations only under the supervision of a master electrician. But there are a couple of major differences in the supervision requirements at this level. For one thing, the one-to-one ratio is not required. One master electrician can supervise any number of apprentice and journeyman electricians, so the one-to-one ratio is not applicable. Moreover, the master electrician is not required to be on the job site at all times, although the expectation is that there will be realistic supervision of the work, including a final review.

Requirements for the New Hampshire journeyman electrician are far more extensive than those for an apprentice card. The initial fee for a three-year license is $150. Each applicant for the *journeyman license* must have acquired 8000 hours of practical or field experience as an apprentice to a journeyman or master electrician while holding an apprentice card. To put this in perspective, if you work a 40-hour week for one year with two weeks off, that would equal 2000 hours, so we are talking about four years of your working life. *Practical experience* is defined as experience acquired while performing electrical installations in a classroom or in the field as part of a course curriculum in an approved school. *Field experience* is defined as experience performing electrical installations in the field under the direction of a master electrician. It does not include time accumulated in a classroom setting. An additional requirement that must be met prior to applying for the journeyman license is 600 hours of electrical schooling, including a minimum of 24 hours on electrical safety, either accomplished in blocks of 150 hours per year, including a minimum of six hours on electrical safety, or by having as associate or higher degree in an electrical curriculum.

Accompanying the journeyman license application, a signed declaration from a New Hampshire master electrician is required, certifying that the applicant has acquired the necessary experience under the supervision of that master electrician.

Many journeyman electricians aspire to becoming master electricians. Here again, this certification is valued by employers. Every electrical job performed by a properly supervised apprentice or by a journeyman electrician must be supervised by a master electrician. A single master electrician is permitted to supervise an electrical job with any number of workers holding journeyman licenses and apprentice cards.

The initial fee for a New Hampshire master electrician license for three years is $270. The applicant must have acquired 2000 hours experience as a journeyman electrician working in the field prior to being allowed to take the examination. A licensed master electrician must sign a document provided by the State of New Hampshire attesting to the applicant's experience.

There is one additional New Hampshire electrician's license. It is neither above nor below the master electrician license but stands apart as a separate category. It is the *high/medium-voltage electrician license*, and it qualifies an individual to work on equipment and circuits that operate in excess of 600 V. The applicant must first obtain an identification card, issued by the New Hampshire Electrician's Licensing Board. Then the applicant must be employed by a high/medium-voltage electrician and obtain certification showing completion of a state, federal, or employer certification program approved by the Board. There is no examination.

Ordinarily, premises wiring does not operate at these elevated voltages, and such work is left to utility linemen. However, some nonresidential facilities will have customer-owned transformers and distribution equipment in which medium and high voltages are present. To legally work on these circuits and equipment in New Hampshire, an electrician must hold the high/medium-voltage license.

Still Struggling

Utility work does not come under *NEC* jurisdiction and does not require an electrician's license, although some utility electrical workers have the license and do private work on the side.

The journeyman and master electrician exams are similar, with one difference noted below. All questions on the *NEC* are based on the current edition. (Some states have been slow to legislate new editions of the *NEC* because they are issued every three years. This presents a dilemma. Should you study the latest edition or some earlier version of it that is enforced in your jurisdiction? Which edition should be brought into an open-book exam? The best thing to do is to contact your jurisdiction's licensing Board to resolve this question.)

The first section of the journeyman and master licensing exams consists of 50 multiple-choice and true-false questions based on the *NEC*. The second section of the exam for both licenses consists of 50 multiple-choice and true-false questions based on practical electrical installations. The third section of the journeyman exam consists of 10 questions based on pertinent state law and administrative rules. The third section of the master electrician exam is similar, but there are 25 questions. Practical questions are based on the *American Electricians Handbook*, published by McGraw-Hill, current edition.

The passing grade is 70 percent on each section of the exam, both for master and journeyman applicants. But here's the hitch: You have to score 70 percent on all three parts. You could score 100 percent on two parts and 68 percent on the third part, and you would not pass. Applicants who do not pass are allowed to retake the exam. Many applicants do not pass the first time. In such cases, it is best to be persistent, analyze your weak areas, study the material some more, and retake the exam. Naturally, it is preferable to pass the exam the first time with as high a score as possible. Applicants who do not pass the first time are required to retake only the section(s) they failed. However, if subsequent attempts are necessary, the entire exam must be taken, and all parts must be passed.

List of Electrical Deficiencies

The New Hampshire Electrician's Licensing Board provides on its website an intriguing page that is titled, "Commonly Found Licensing Concerns and Installation Deficiencies." The most recent version is titled "2011 Deficiencies." (Similar documents from previous years are archived.) The Board comments at length on these electrical deficiencies frequently found by New Hampshire electrical inspectors in the course of onsite visits:

- The improper installation and securing of expansion fittings in runs of rigid polyvinyl chloride (PVC) conduit
- The use of electrical equipment without following the manufacturer's instructions

- Tightening of electrical connections
- Clearances: working space, clear space, headroom, and dedicated equipment space
- Identifying grounded and equipment-grounding conductors
- The identifying of ungrounded branch-circuit conductors when there is more than one nominal voltage system on the premises
- The sealing of underground raceways
- The bonding of metal water piping in the vicinity of separately derived systems
- Wireways
- The use of NM cable in other structures permitted to be Type III, IV, or V construction
- The improper installation of exposed vertical risers from fixed equipment
- The mounting height of switches
- Circuit directories or circuit identification
- The terminating of more than one grounded conductor under a single terminal
- The improper connection of septic pumps
- The flexible connection to emergency-system equipment in hospitals
- The continued failure to install signs that are listed by a recognized third-party testing agency
- Emergency, legally required, and optional standby signs
- The separation of emergency-system conductors from other conductors
- The location of the disconnecting means for emergency systems
- The installation of branch circuits supplying emergency lighting

Still Struggling

The list of frequently found deficiencies in electrical installations is made up of *NEC* violations. This document, put out by the National Fire Protection Association (NFPA), governs most nonutility electrical installations. A new edition is issued every three years. The first edition appeared in 1897.

Getting Certified in Oregon

Turning now to Oregon, another state with a well-organized and effective regulatory infrastructure for electrical workers, we shall examine its user-friendly website and see what we can learn from it.

Like New Hampshire, Oregon has a highly developed regulatory structure for electrical workers within the state. There are some significant differences, but the underlying concepts are similar. The Electricians' Board comes under a larger regulatory agency, the Building Codes Division (BCD), which licenses workers in the boiler and pressure-vessel, electrical, plumbing, and manufactured-housing industries.

Oregon recognizes two principal types of electricians, journeyman and supervising. (Supervising electricians are similar to New Hampshire's master electricians.) The requirements for a *journeyman license* are one of the following:

- Completion of an Oregon-approved apprenticeship program
- Completion of 576 hours of classroom training and 8000 hours of work experience outside Oregon (This must include at least 1000 hours in each of the following: residential, commercial, and industrial.)
- Completion of 16,000 hours of work experience outside Oregon with a minimum of 2000 hours in each of the following: residential, commercial, and industrial

An exam is given by the Oregon Licensing Board. It consists of 52 questions with a time limit of three hours. The passing grade is 75 percent. The examination fee is $10. The license application and renewal fees are $100, good for three years.

The Oregon *general supervising electrician license* permits the holder to perform any electrical installation when employed either by a licensed electrical contractor or by an industrial plant. There is a $100 application fee, which includes the first three-year cycle. Thereafter, the renewal fee is $100 for each three-year period.

The exam is made up of two sections. Section one contains 52 questions, and section two contains 12 questions. There is a four-hour time limit. The exam is open-book, and the following materials are allowed:

- Current edition of the *NEC*
- Current *Oregon Electrical Specialty Code* (available as a free Internet download)

- Current *NEC Handbook*
- *Tom Henry Key Word Index*
- *Ferm's Fast Finder Index*
- *Ugly's Reference*
- *American Electricians Handbook* (published by McGraw-Hill)
- *Oregon Administrative Rules*
- *Oregon Revised Statutes*

QUIZ

These questions are intended to test your comprehension of Chapter 1. The passing score is 70 percent, but try to answer them all correctly. The quiz, like most electricians' tests, is open-book, so feel free to refer to the text. Answers appear in Answers to Quizzes and *NEC* Practice Exam.

1. **All states have jurisdiction over the licensing of electricians.**
   A. True
   B. False

2. **To find out what materials you may bring into the electrician's licensing exam,**
   A. consult the *NEC*.
   B. consult the Electricians' Board website.
   C. call the Electrician's Board.
   D. B or C.

3. **In New Hampshire, satellite dish installation**
   A. does not require an electrician's license.
   B. is tightly regulated.
   C. is considered power and light work.
   D. does not require grounding.

4. **In many jurisdictions, home owners may do installations**
   A. for neighbors, not for pay.
   B. as long as they don't tie into the entrance panel.
   C. and it is not required to comply with the *NEC*.
   D. on their primary residence only.

5. **In New Hampshire, the apprentice card permits an individual to**
   A. work as an independent electrician.
   B. work under the supervision of a journeyman electrician with a master electrician overseeing the job.
   C. work on low-voltage systems only.
   D. install wiring that is not tied into the entrance panel.

6. **In New Hampshire, the high/medium-voltage license is required to work on circuits over**
   A. 120 V.
   B. 240 V.
   C. 600 V.
   D. 1800 V.

7. **Utility work does not come under *NEC* jurisdiction.**
   A. True
   B. False

8. **Frequently seen electrical deficiencies involve**
   A. the use of electrical equipment without following the manufacturer's instructions.
   B. the tightening of electrical connections.
   C. the sealing of underground raceways.
   D. All of the above

9. **Which of the following written materials may be brought into the Oregon electricians' licensing exam?**
   A. Current edition of the *NEC*
   B. Current *NEC Handbook*
   C. *Ferm's Fast Finder Index*
   D. Any of the above

10. **Improper connection of septic pumps is a concern of electrical inspectors.**
    A. True
    B. False

chapter 2

The National Electrical Code

Elementary school students learn that Thomas Edison invented the light bulb. Actually, his work went well beyond that. He and his many associates designed and built an electrical generation and distribution system for lower Manhattan that brought electrical power to numerous residential and commercial customers. Basic concepts, such as overcurrent protection by means of fusing and even circuit breakers, had been invented decades earlier for use in telegraph systems. This was a good start, but far from perfect. As may be expected, there were a great many fatalities due to electrical fires and shock. Insurance companies sustained great losses, especially in the 1880s as electrical distribution systems expanded and usage became more widespread. In an attempt to remedy this situation, many electrical codes surfaced in the United States and Europe. Naturally, these standards had differing, often conflicting provisions. Many problems remained unresolved, particularly with regard to grounding. Some standards mandated grounding, whereas others made it optional, and in a few regions it was outrightly prohibited. (Grounding facilitates the operation of overcurrent devices, but as practiced at that time, it could set the stage for increased electric shock hazard.)

CHAPTER OBJECTIVES

In this chapter, you will

- Learn about the origins and early development of the *NEC*.
- See what is covered and what is not covered in the *NEC*.
- Discover a method for navigating quickly through the *NEC*.

TIP ***Don't try to memorize the NEC. Learn where the requirements are located so that you can find them quickly during an open-book exam.***

In 1896, concerned individuals representing a number of organizations met in New York City and drafted a set of standards intended to resolve the conflicts mentioned earlier. After much discussion and review, the first *National Electrical Code* (*NEC*) was published in 1897. Revisions have been issued periodically—since 1975, once every three years. There are numerous changes in each edition, some consisting of a single altered word or phrase, others adding entire articles. The trend is toward more exacting requirements, although in some cases mandates are eliminated if it is determined that the safety of the electrical installations will not be compromised.

NEC and the Jurisdiction

At the back of the current *NEC* are three copies of a form that any interested person can use to propose a change. All such proposals are considered by one of the code-making panels. The proposals are released for public comment and put to a vote by the National Fire Protection Association (NFPA) at its next annual meeting.

As enacted by the NFPA, the *NEC* has no legal standing. By publishing it, NFPA is making it available so that any jurisdiction can legislate it into law in whole or in part, with or without amendments. States, municipalities, housing authorities, insurance companies, military installations, and entire countries have enacted it. Besides most parts of the United States, the *NEC* has been adopted by Mexico, Costa Rica, Venezuela, and Columbia. It has been translated into Spanish, Korean, and Japanese.

Canada has its own regulatory document, the *Canadian Electrical Code* (*CEC*), whereas European Union (EU) countries recognize the International Electrotechnical Commission (IEC) mandates. The trend seems to be in the direction of an international standard that individual countries could choose to enact.

NFPA publishes other codes and standards. Electricians who do advanced design and construction are familiar with those that affect their work. For example, fire alarm systems are quite complex. Designers and technicians frequently refer to these NFPA documents:

- NFPA 101, *Life Safety Code*, lists the types of buildings that are required to have fire alarm systems.

- NFPA 72, *National Fire Alarm Code*, lists design specifications, including spacing and location of heads, pull stations, horns, strobe light alarms, and the like. Also included are maintenance and testing procedures, operational protocols, and performance standards.
- NFPA 70, *National Electrical Code*, Article 760, "Fire Alarm Systems," covers wiring and equipment details. This includes power to the control console, zone wiring initiating devices and indicating appliances, and phone lines used for automatic calling to the local fire department, monitoring agency, or any other numbers that may be programmed into the control panel. If controlled by the fire alarm system, the *NEC* also covers other fire alarm functions, including sprinkler water flow, sprinkler supervisory equipment, guard's tour, elevator capture and shutdown, door release, smoke door and damper control, fire doors and fan shutdown, and liquid petroleum gas supply. Article 725, "Class 1, Class 2 and Class 3 Remote Control, Signaling and Power-Limited Circuits," focuses on wiring emanating from the control panel if the system is power-limited. Included are alternative requirements for minimum wire sizes, derating factors, overcurrent protection, insulation, and wiring methods and materials.
- Underwriters Laboratories (UL) and similar agencies list all fire alarm system components, including control panel, smoke-detecting heads, pull stations, and batteries.
- NFPA 70E, *Standard for Electrical Safety in the Workplace*, specifies on-the-job safeguards and work procedures.

"Informational Notes" occur throughout the *NEC*. Unlike the main text, they are explanatory or advisory and are not enforceable mandates. They do not include the words *shall* or *shall not*. (They used to be called "Fine Print Notes").

What Is Not Covered

The *NEC* limits its own jurisdiction by excluding certain areas: ships other than floating buildings; railroad rolling stock, aircraft, and automotive vehicles other than mobile homes and recreational vehicles. Underground installations in mines are also not covered, but this exclusion does not involve nonmine below-grade wiring such as buried electrical lines or lighting in a traffic tunnel.

Still Struggling

Utility lines do not come under *NEC* jurisdiction, but they are covered by the *National Electrical Safety Code*, also issued by NFPA. Included in the mandates are required heights above grade for power lines of various voltages, what occupancies are required to have fire alarm systems, and other safety-related matters.

Also excluded are communications and power utilities equipment directly involved in generation, distribution, and transmission. Office space, machine rooms, employee lunchrooms, and truck maintenance garages owned and/or operated by a utility are covered in the *NEC*.

Frequently the Code mentions the *authority having jurisdiction* (AHJ). This individual is generally an electrical inspector. The AHJ is charged with the task of interpreting the Code rules and decides whether or not an installation is in compliance. As part of the work, the AHJ decides whether electrical equipment is safe and may be used. Neither the Code nor the AHJ looks inside factory-made equipment such as an electric stove or a computer. The AHJ does not have the equipment or knowledge to examine the inner workings of these items. Instead, the AHJ accepts or rejects the equipment as a whole based on whether or not it is listed by a testing organization such as UL. These outfits have extensive testing facilities and established procedures that permit them to rule on the suitability of the equipment from a safety standpoint. In recent years, counterfeit listing has become a problem. In purchasing electrical tools and materials, you can protect yourself from hazards resulting from installing unsafe equipment by buying from reputable dealers. If the listing is genuine, it can be assumed that a product will meet minimum safety standards, provided that it is installed according to the manufacturer's instructions, which accompany the product and are part of the listing.

The *NEC* does not contain design specifications, and it is not a manual for untrained persons. It is a highly specialized document. The purpose is the practical safeguarding of persons and property from hazards arising from the use of electricity. The two principal hazards are fire and electric shock. There could be other hazards as well. For example, an inadequately secured conduit could fall and injure a person below.

Potential Hazards

Far more fatalities are caused by electrical fire than by electric shock, especially if utility work–related events are excluded. And most fire victims perish due to smoke inhalation as opposed to burning. These unfortunate matters must be considered carefully when wiring any building. You should always think of the child who may be sleeping on the top floor of a building you have wired.

PROBLEM 2-1

What are the two major hazards arising from the use of electricity?

SOLUTION

They are electrical fires and injury from electric shock.

In mitigating hazards arising from the use of electricity, by far the most important tool is the *NEC*. If your wiring is Code-compliant in every detail, you can be sure that you are not building hazards into your work.

Ground-fault circuit interrupters (GFCIs), now required in sensitive locations in dwellings and nondwellings alike, have cut way down on the number of electrocutions that occur annually. And the newer arc-fault technology holds great promise for abating the scourge of electrical fires.

As you become more familiar with the *NEC*, you will learn its peculiar use of language. For example, in a licensing exam or on the job site, you may want to find the required depth for an underground electrical line, type of raceway, and occupancy. Within the index entry for underground wiring, burial depth is not mentioned. The Code location is found under "Minimum Cover Requirements." You go to that section, which refers to the relevant table.

TIP ***If you are preparing for a licensing exam, become very familiar with the NEC Index and Table of Contents. In an open-book exam, one or the other of these will take you right to the answer.***

There are pages within the Code containing information that will never appear on an exam. An example is a series of pages before and after the Contents and before the Introduction. In some editions, these pages include extensive listings of members of Code-making panels and similar nonessential information. These pages may be folded over and taped so that they don't impede your efforts during an exam.

Also, some of the tables contain lots of information that is outside our main focus. When exam questions involve ampacity calculations, it is generally assumed that they are referring to copper, unless aluminum is specifically mentioned. The same applies to the Fahrenheit as opposed to Celsius temperature scales, except in the case of conductor insulation temperature ratings, where metric figures are customarily used. To simplify, you can draw light pencil lines through the less relevant portions of the tables and through some wording in the text. This will help you to move faster during the exam. But such markings in your Code book should not be made before clarifying with your local electricians' board because it could be construed as bringing notes into the exam.

QUIZ

These questions are intended to test your comprehension of Chapter 2. The passing score is 70 percent, but try to answer them all correctly. The quiz, like most electricians' tests, is open-book, so feel free to refer to the text. Answers appear in Answers to Quizzes and *NEC* Practice Exam.

1. Any interested person can propose a change to the NEC.

A. True
B. False

2. The *NEC* has been translated into

A. French, Russian, and Chinese.
B. Spanish, Korean, and Japanese.
C. Chinese, Greek, and Norwegian.
D. All of the above

3. Besides *NEC*, another electrical code is

A. *CEC*.
B. *IEC*.
C. Both of these
D. Neither of these

4. NFPA publishes

A. the *National Electrical Code*.
B. the *Life Safety Code*.
C. the *National Fire Alarm Code*.
D. All of the above

5. *NEC* does not have jurisdiction in

A. municipal buildings.
B. railroad rolling stock.
C. mobile homes.
D. underground traffic tunnels.

6. The AHJ

A. interprets Code rules.
B. inspects internal circuitry in factory-made equipment.
C. checks utility wiring crossing private property.
D. administers the licensing of electricians.

7. Most fire victims perish due to

A. burning.
B. smoke inhalation.
C. physical exhaustion.
D. None of these

8. GFCIs are not required in

A. private homes.
B. commercial locations.
C. factories.
D. They are required in sensitive areas within all the preceding locations.

9. When exam questions refer to conductors, it is assumed that copper is meant unless stated otherwise.

A. True
B. False

10. To find answers to a question in the *NEC*, a good method is to

A. consult the Index.
B. consult the Table of Contents.
C. page through the main text.
D. A or B

chapter 3

Preparing for the Exam: What to Expect (It's Not as Hard as You Think)

In most states and jurisdictions where an electrician's license is required, an examination is the heart of the matter. It is difficult but not as bad as you may think. True, a good portion of the applicants do not pass the first time. But most electricians' boards allow those who fail the exam to retake it, with the prospect of eventually passing. Many are woefully unprepared. Believing that the test will be a mere formality, they are aghast at the amount of knowledge required to find the answers on an open-book basis given the time constraints. But the fact that you are reading this book indicates that you are motivated and that you will pass the exam first time around.

CHAPTER OBJECTIVES

In this chapter, you will

- Find out how to prepare for an electrician's licensing exam.
- Learn how to consult the *National Electrical Code* (*NEC*) for each new job.
- See how best to use the *NEC* Table of Contents and Index.

To begin, consult your electricians' board website and gather whatever information is available. Most licensing exams are open-book tests. Find out what books and materials you are allowed to bring into the exam. Some jurisdictions allow only the *National Electrical Code* (*NEC*) with no notes, whereas others permit additional books. The *NEC* Index Tabs are a great time-saver and are highly recommended, but some jurisdictions do not allow them.

Still Struggling

You will find the answers to procedural questions on your jurisdiction's website. In the search engine, type the name of your state or municipality and "Electricians' Licensing Board," and that will bring up what you need.

Be Prepared

Most jurisdictions allow a handheld calculator, but they exclude specialized electrical code calculators, scientific calculators, and laptops. You don't need anything elaborate with multiple memory levels—just an inexpensive handheld calculator with a readable display. If it can add, subtract, multiply, divide, and do squares and square roots, that is all you need.

TIP ***Now is the time to get started. Find out what is needed to get an apprentice permit in your jurisdiction. Contact electrical construction firms in your area to see if you can get on the job so that you can begin accumulating hours.***

Preparing for the exam is best accomplished on multiple fronts. Assuming that you have acquired an apprentice permit and that you have begun working in that capacity, there are several actions you can take that will enhance your ability to score well on the exam.

Depending on your prior knowledge and experience, your journeyman or master electrician supervisor probably will ask you to perform some relatively simple but nevertheless important tasks. If it is a residential job, you may spend your first day drilling holes in studs or running Romex cable between wall boxes and doing "home runs" back to the entrance panel. The next day's assignment could be laying polyvinyl chloride (PVC) conduit with a pull rope in a ditch and installing sweeps, expansion joints, and connectors at either end. Apprentices who show aptitude, do quality work, and get a lot done in a day are usually reassigned to more advanced work, such as putting together a complete service or wiring an entrance panel.

The point of all this is that while you are performing this work, you can be preparing for the journeyman and master exams that you will eventually take.

PROBLEM 3-1

Why is it important to get an apprentice card and begin work?

SOLUTION

To qualify for a journeyman license, you need to acquire hours working in the field under the supervision of a master electrician who will certify your hours.

You should have a copy of the *NEC* with you on the job and refer to it whenever you venture into new territory. If you are running Romex, consult Article 334, "Nonmetallic-Sheathed Cable." Referring to "Uses Permitted" and "Uses Not Permitted," verify that the cable is the correct choice for the application. Check on the securing intervals, minimum radius of bends, and other installation details.

Keeping a Journal

It is suggested that right from the start you keep a journal with entries for every job you do. Include Code references with chapter and article numbers. When it comes time to take the exam, you will be able to think back and determine the correct answers to many questions or, alternately, know exactly where to look in the *NEC*.

In addition to the Chapter 3 articles on each cable and raceway type, there are numerous other Code entries that are relevant. If you refer to ampacity and conduit fill tables that affect your work, they will become very familiar, and when you take the exam, you will know precisely where to find the needed information. If you are fortunate enough to have a supervising journeyman or master electrician who is willing to share information and discuss Code interpretations as opposed to merely putting in the time and getting the job done, you can make note of this material in your journal, and it will come back to you when needed on the licensing exam.

If you really want to reinforce this knowledge, when you get home at night, spend a few minutes typing this information into a file on your computer, and back it up so that it will survive in the hard drive.

In the time it takes to acquire four years of work experience or whatever your jurisdiction requires, it is likely that you will amass experience relevant to a good proportion of the questions on your licensing exam. This is one technique that will lead to a good exam score. There are others as well. The thing to keep in mind is that most of us can never learn and retain all the material in the *NEC*. Fortunately, the exam in most jurisdictions is an open-book test, and this is realistic because on the job site or in the shop, you can and will be referring to the Code when questions arise. Don't waste your time trying to memorize all the detailed information that is contained in the *NEC* text and tables. Instead, learn the overall structure of the Code so that you can quickly find the correct answer.

Timing Is Everything

If there are 60 questions on the *NEC* section of your exam and you have two hours to complete it, this means that you have two minutes to spend on each question. You will perhaps instantly know the answers to some of the questions, so let us say that this leaves an average of three minutes for each of the questions that have to be researched. What this means is that you need to find the answers rather quickly as opposed to paging through the text in the hope of finding the correct location. This is called *Code hopping*, and it will ensure a low score on the exam.

Answers to some questions are easy to find. For example, if you are asked a multiple-choice question that involves the definition of *dwelling unit*, you will know that it is alphabetized within Article 100, "Definitions." If you have Code tabs, you can flip there instantly. Other answers may be more difficult to find. Tap rules are spread throughout the Code depending on whether you are talking about feeder taps, motor taps, or whatever. If you have three minutes to spend on a question and no idea where to look, this will require a different approach.

It is this type of Code learning, *structural analysis*, that has to be assimilated—certainly a manageable task that is far easier than memorizing every individual requirement. Get used to using the Index and Table of Contents.

Some Code locations are found more easily using one of them, and some are found more easily using the other.

If your state permits the use of *Ferm's Fast Finder*, it is highly recommended. Lots of times it will get your answer, and you will still be under the three-minute allotment. But you should not rely on it for every question or you will lose time that way.

You may know the answer right off the top of your head. If not, start with the Code book. If possible, with the help of the Index tabs, go to the relevant section and find your answer. Otherwise, consult either the Table of Contents or the Index. (Before entering the exam room, you should try to become an expert on use of the Table of Contents as opposed to Index usage. If you know which of them to use for various subjects, you're on track to get the right answers quickly.)

If this procedure doesn't seem like its going to get the right answer for you, turn to *Ferm's Fast Finder* and go from there.

Given unlimited time, you could find the correct answer to every question, but if you have to spend 15 minutes researching a single question, that's going to be disastrous for the rest of the exam.

QUIZ

These questions are intended to test your comprehension of Chapter 3. The passing score is 70 percent, but try to answer them all correctly. The quiz, like most electricians' tests, is open-book, so feel free to refer to the text. Answers appear in Answers to Quizzes and *NEC* Practice Exam.

1. Information concerning electricians' licensing in individuals states is

A. compiled in the *NEC*.
B. available on the electricians' board website.
C. a closely held secret.
D. uniform throughout the United States.

2. Most jurisdictions do not permit which of the following in exams?

A. The *NEC*
B. Handheld calculators
C. Answer keys
D. All of the above

3. To get an electrician's license, in most jurisdictions you have to start with an apprentice card.

A. True
B. False

4. To pass the electricians' licensing exam,

A. you need to memorize the *NEC*.
B. it is best to bring your lawyer.
C. you need to work quickly.
D. is a nearly impossible task.

5. All states use the same exam.

A. True
B. False

6. To find answers in the Code, which of the following is(are) useful?

a. Table of Contents
b. Index
c. Code tabs
d. All of the above

7. Ferm's Fast Finder is

A. permitted in all exams.
B. permitted in some exams.
C. never permitted.
D. sold in all bookstores.

8. Code tabs

A. are useless in the exam.
B. will guarantee that you pass.
C. are permitted in some states.
D. are permitted in all states.

9. An apprentice

A. is permitted to do electrical installations with supervision.
B. must first pass a rigorous examination.
C. must have a degree in electrical engineering.
D. All of the above

10. The definition of *dwelling* appears in

A. the Code Annexes.
B. Article 100.
C. any good dictionary.
D. each of the relevant articles.

chapter 4

Multiple-Choice Strategy: Using Logic to Improve Your Score

Most electricians' licensing exams combine multiple-choice and true-false questions. These tests are easier to score than those with open-ended questions, which are often subjective. Generally, an answer sheet is provided, and you pencil-in the space provided. These answer sheets can be graded electronically because the pencil marks are conductive. If you need to change an answer, X it out or thoroughly erase the wrong answer depending on instructions. Some examinations, especially those provided by independent contractors, involve the use of desktop computers at the exam site. You will not have a problem if you need to change an answer.

CHAPTER OBJECTIVES

In this chapter, you will

- See how to maximize your test score.
- Learn about some common mistakes that are made interpreting exam questions.

Pay close attention to the instructions. Generally, a passing grade is 70 or 75 percent. You are graded on the number of correct answers. Therefore, you do not want to leave any questions unanswered, even if you do not know the correct answer. If you are running out of time, quickly go through any unanswered questions and give them your best shot.

TIP ***Going into the exam, make sure that you have the materials permitted. Bring spare batteries for your handheld calculator and, if permitted by your electricians' licensing board, Code tabs for your copy of the National Electrical Code (NEC).***

Multiple-Choice Alternatives

Multiple-choice questions usually have four alternate answers lettered A through D. Generally, it is specified that you are to choose the best answer. It is possible that a choice might in some technical sense be correct but still not the *best* answer. Don't get into scholastic debates. Choose the single best answer, not more than one. If there were four choices, you could figure on getting a 25 percent score if you had no knowledge of the subject and guessed at every question. True-false questions, based on a total lack of knowledge, would score 50 percent.

The best way to proceed is to quickly scan down through all the questions. The wording of some questions may provide answers or at least eliminate some choices for other questions. Also, in seeking an answer in the Code, you may come across the answer to a different question, so it is a good idea to read through all the questions at the outset if you can do that quickly.

Still Struggling

Find as much information as possible about the exam format, contents, and so on from the electricians' board website. Most exams conform to a single pattern—multiple-choice or true-false. On the Internet, there are numerous sites with practice questions that are scored electronically.

Using Logic

You will undoubtedly know the answers to some of the questions with a high degree of certainty. Answer those first, and then answer any that require you to consult the Code if you know right where to look. Often you can eliminate at least one of the answers for sure, and sometimes you can narrow it down to a choice between two alternatives that make sense. In many cases, two of the answers exclude each other. An example is this familiar combination:

C. All of the above.

D. None of the above.

Because these cannot both be true, you have an advantage in this type of question. But these choices are not all-encompassing, so it cannot be said that any choices have been eliminated. The situation is different when two choices are mutually exclusive and all-inclusive, as in the following example:

A. AWG 6 or larger.

B. Smaller than AWG 6.

In this example, the answer usually will be A or B, but not always because while A or B will usually be true, it may not be the *best* answer.

Be sure to distinguish between "AWG 6 or larger" and "Larger than 6 AWG" because the latter does not include 6 AWG, so it is not all-inclusive along with B.

PROBLEM 4-1

How is it possible to maximize the test score?

SOLUTION

You need to learn how to navigate the Code quickly so that you do not run over the time limit. Additionally, start thinking about the logic in multiple-choice questions. Sometimes you can eliminate alternatives because of the way they are worded.

Read Carefully

Many Code requirements apply to systems that operate at over 600 V, so the nominal 600-V system is not included. These requirements pertain to nominal values, even if the measured voltage is a little higher or lower.

This is the sort of language that you have to watch carefully. In so doing, you can eliminate or narrow down alternate choices.

It is easy to neglect a negative that changes the whole meaning of a question. *Not* and *never* are frequently overlooked, the unfortunate outcome being that there is an incorrect answer. *Minimum* and *maximum* are also confused. Think about each question, and make sure that none of these small words are misconstrued. It is possible to know the Code material quite well from a technical point of view but make mistakes in the wording and as a consequence score low. Perhaps we all are neglectful in varying degrees. I know that I am, and I have to watch the wording carefully. If you practice scrutinizing each question carefully to make sure that it is not misinterpreted, you will make big improvements.

Before taking an exam, it is good to do practice questions. You can analyze the results to see whether you have weak areas that need work. At the back of this book is a practice exam with answers. Follow the instructions, and see if you score 75 percent or better. If so, you should pass the licensing exam. For any questions you missed, go back to the *NEC* and research the topic thoroughly.

QUIZ

These questions are intended to test your comprehension of Chapter 4. The passing score is 70 percent, but try to answer them all correctly. The quiz, like most electricians' tests, is open-book, so feel free to refer to the text. Answers appear in Answers to Quizzes and *NEC* Practice Exam.

1. **Electricians' licensing exams contain only true-false questions.**
   A. True
   B. False

2. **In an electricians' licensing exam,**
   A. unanswered questions are not counted.
   B. do not leave any questions unanswered.
   C. there is no hurry, and you can take as long as you want.
   D. if you are unsure of your ability to pass the test, you can hire someone to take it for you.

3. **Code tabs**
   A. are never allowed in exams.
   B. may be misleading.
   C. are an excellent time saver.
   D. are required in most exams.

4. **In a licensing exam, there is a time limit, so you have to hurry.**
   A. True
   B. False

5. **If an electrical system is 600 V nominal but it measures 605 V,**
   A. it should be considered over 600 V just to be safe.
   B. it is considered not over 600 V.
   C. it could go either way.
   D. it doesn't matter because Code requirements are the same.

6. **Practice exams are beneficial because**
   A. they highlight weak areas.
   B. they help you to learn to use the Code book efficiently.
   C. they increase your reasoning ability.
   D. All of the above

7. **In an exam, you should take as much time as necessary to get every answer correct.**
   A. True
   B. False

8. If you do not pass a licensing exam,

A. you will never be able to retake it.
B. you will be banned in all jurisdictions.
C. you are not alone. Study your weak areas, and try again.
D. you need to find a new profession.

9. The *National Electrical Code*

A. is the only book allowed in all licensing exams.
B. is the only book allowed in some licensing exams.
C. has few changes, so you don't need the current edition.
D. is not very useful because it is overly technical.

10. If electricians didn't have to be licensed,

A. the world would be better off.
B. there would be fewer electrical fires.
C. electricians would be more highly paid.
D. None of the above

chapter 5

National Electrical Code *Chapters 1 through 3*

A good working knowledge of the *National Electrical Code* (*NEC*) is essential for passing electricians' licensing exams and successfully completing on-the-job design and installation work. As noted previously, you will never be expected to know all the thousands of provisions contained in text and tables of the *NEC*. Instead, you need to be able to look up this information quickly. This is one of the very basic skills absolutely necessary in the electrician's trade.

CHAPTER OBJECTIVES

In this chapter, you will

- Survey basic concepts for residential and commercial wiring.
- Examine electricians' methods for wiring protection.
- Learn about *NEC* sizing tables for conductors and raceways.

In order to locate information in the *NEC*, you need to understand its structure and become familiar with its unique use of language. We'll take a close look at these subjects in this chapter and the ones that follow. At the end of this book are multiple-choice and true-false questions and answers. They are the same level of difficulty as you will encounter on a licensing exam. And like most licensing exams, these tests are intended to be taken on a timed, open-book basis. If you can consistently score 70 percent or better, you should pass the *NEC* section of your licensing exam. If not, go back through your incorrect answers to pinpoint weaknesses so that you can improve your Code navigation and/or test-taking ability. It is to be emphasized that you need to have your copy of the *NEC* in order to do the practice exam as well as to participate in the licensing process. Now is the time to order it. The *NEC* is available at a discount at Amazon.com.

Scoring Well

Now we'll begin our survey of the *NEC*, starting with the Introduction and first three chapters. What you should remember is that the *NEC* begins with general principles and goes on to the specific. In other words, any provision in the first three chapters pertains to equipment, installations, and working methods covered in the later chapters unless specifically modified or overruled. An exception to this is Chapter 8, "Communications Systems," where the general principles in Chapters 1 through 3 are not applicable unless specifically referenced. This distinction might seem fanciful, but it could make a big difference in how an electrical installation is put together.

Large amounts of memorization are not necessary to score well on the licensing exam. However, I suggest that you do memorize the title of each chapter because this will greatly increase your speed and accuracy in navigating the Code.

I said that the Code starts with the general and moves toward the specific. The most general part of the entire document, then, is Article 90, "Introduction." This brief section lays out the underlying assumptions that govern the way the entire document works. Licensing exams generally have questions that focus on the "Introduction." You will know the answer immediately, or the Code location will be evident and you will easily find the answer. The "Introduction" opens with a statement of purpose, which is the practical safeguarding of persons and property from hazards arising from the use of electricity. Injury to persons is primarily due to shock and fire. Damage to property includes not only the building and contents but also damage to the wiring and electrical equipment. The Code makes the point that compliance will result in an installation free of hazards but not necessarily efficient, convenient, or adequate for good service or future expansion of electrical use. Moreover, it is stated that the Code is not intended as a design specification or manual for untrained persons.

Thus we see that the Code carefully limits its own purpose and scope. It covers the installation of electrical conductors, equipment, and raceways, including communications and optical fiber installations, for the following:

- Public and private premises, including buildings, structures, mobile homes, recreational vehicles, and floating buildings
- Yards, lots, parking lots, carnivals, and electrical substations
- Installations of conductors and equipment that connect to the supply of electricity
- Installations used by the electric utility, such as office buildings, warehouses, garages, machine shops, and recreational buildings, that are not an integral part of a generating plant, substation, or control center

The "Introduction" concludes with Section 90.9, "Units of Measurement." It discusses the status of the two systems of units—metric and U.S. customary units. In keeping with the international shift toward greater use of metric units, current editions of the Code provide metric figures first, followed by U.S. customary units in parentheses. However, licensing exams generally ask questions and expect answers in inch-type measurements only.

Definitions and Requirements

Chapter 1, "General," consists of only two articles. They are Article 100, "Definitions," and Article 110, "Requirements for Electrical Installations." Answers to questions concerning definitions are very easy to locate. The article is simply an alphabetical listing of terms. Article 110 is a little more difficult because rather than a list of definitions, it contains requirements for all electrical installations, subject to the modifications we discussed earlier. Some of the more important topics are

- Conductor sizes
- Wiring integrity
- Interrupting rating
- Electrical connections and temperature limitations
- High-leg marking

- Arc-flash hazard warning
- Available fault current
- Spaces about electrical equipment

This is not to say that there are not other provisions in Article 110, but the topics just listed are very likely to appear on the exam and in a job-site installation. Electrical inspectors check for compliance (Figure 5-1).

Chapter 2, "Wiring and Protection," is nearly four times as long as Chapter 1. It is also very general, although a trend toward the specific will be discerned. The most important articles cover

- Use and identification of grounded conductors
- Branch circuits
- Feeders
- Branch-circuit, feeder, and service calculations
- Outside branch circuits and feeders
- Services
- Overcurrent protection
- Grounding and bonding

These articles comprise an essential part of the Code, and they are applicable to almost every wiring job. They are lengthy and quite complex, especially Article 250, "Grounding and Bonding," which is longer than some whole chapters in the Code. Article 220, "Branch-Circuit, Feeder and Service Calculations," is very frequently referenced, and it is the focus of many exam questions. You will be

FIGURE 5-1 • Insulating bushings used to protect conductors where they emerge from a connector.

asked to calculate conductor sizes using rules and procedures found in Article 220. Thanks to the intrepid number-crunching ability of the handheld calculator, the math is easy. But you will have to know the anatomy of Article 220 so that you can decide how to factor in the many parameters involved in various occupancies.

Because of the fundamental importance of Chapter 2, we'll discuss the individual articles.

Article 200, "Use and Identification of Grounded Conductors," covers requirements for grounded conductors in premises wiring, identification of terminals, and identification of grounded conductors. Of course, conductors and terminals have to be properly identified to prevent cross-wiring, which would constitute a shock hazard.

First, we have to distinguish between a ground*ed* conductor and a ground*ing* conductor. Despite the fact that both are usually at the same potential, 0 V with respect to an ideal zero-impedance grounding electrode; nevertheless, the two wires serve different purposes and are terminated and identified differently. Always pay attention to whether the word ends in *–ed* or *–ing*. The purpose of the grounded conductor is to complete the circuit, providing a path for the current to return to the power source. The grounding conductor, on the other hand, does not normally conduct current. At the load, it is connected to any current-carrying conductive part, such as the metal housing of a power tool that is not double-insulated. It is also electrically connected, through connectors at either end, to any metal raceway and metal enclosures. At the service-entrance panel, it is connected to the grounding bar, which is mounted on the inside of the metal enclosure and bonded to it.

Separate Paths

The grounded conductor, in contrast, is connected to the neutral bar, which has a similar appearance. This grounded bar is electrically isolated from the enclosure unless it is a service-entrance panel, in which case a main bonding jumper is installed. This piece of hardware usually takes the form of a long-threaded bolt that is screwed through the grounded terminal, digging into the metal enclosure to create the necessary electrical connection between the grounded conductor and the grounding conductor. Downstream, they must be kept separate, never to rejoin. If they were to be connected anywhere downstream, including at the load, there would be dangerous circulating currents due to inevitably unbalanced loads.

Still Struggling

The equipment-grounding conductor is always required. The circuit may appear to work without it, but it should not be omitted because it provides essential protection against shock and fire hazards.

Keeping these conductors separate depends on having them properly identified. To this end, the Code has various provisions both for conductors and for terminals. Grounded conductor sizes 6 American Wire Gauge (AWG) or smaller are to be identified by one of the following:

- A continuous white outer finish
- A continuous gray outer finish
- Three continuous white stripes along the conductor's entire length on other than green insulation
- Wires that have their outer covering finished to show a white or gray color but have colored tracer thread in the braid may be used as grounded conductors.

Notice that 6 AWG wires are included in the smaller size. If the grounded conductor is 4 AWG or larger, it may be identified as earlier like the smaller wires, or it may be identified at the time of installation by a distinctive white or gray marking at its terminations. This could be paint, tape, or some other means. A product that resembles black electrical tape is available in different colors, including white. This is called *phase tape*, and it is the most convenient way to reidentify conductors. The marking must completely encircle the conductor. Many electricians make three rings, but just one is compliant. This field marking is the most frequent method for identifying the grounded conductor if it is 4 AWG or larger. This is so because it is not feasible to stock multiple reels of large conductors in different colors.

TIP ***Color-coding, even where not required, is a big help in keeping your work organized and will facilitate maintenance and troubleshooting in the future.***

In addition to conductors, terminals must be identified so as to distinguish the grounded pole. Article 200 provides that the identification of terminals to which a grounded conductor is to be connected must be substantially white in color, and the identification of other terminals is to be a readily distinguishable different color.

The concluding section of Article 200 states by way of summary that no grounded conductor is to be attached to any terminal or lead so as to reverse the designated polarity.

Branch Circuits

Article 210, "Branch Circuits," is fairly lengthy and should be studied carefully because it applies to many electrical installations. Trunk slammers make many errors, especially in residential wiring, that are a result of inattention to this article. Licensing exams focus on this material, so a close reading of it is in order (Figure 5-2). Section 210.8, "Ground-Fault Circuit-Interrupter Protection for Personnel," should be scrupulously observed to ensure that this lifesaving technology is in place (Figure 5-3). Ground-fault circuit interrupters (GFCIs) work by comparing the amount of current on the ungrounded (hot) conductor with the amount of current on the grounded (neutral) conductor. If the difference exceeds 4 to 6 mA, indicating a ground fault, the device cuts out, greatly limiting shock hazards. There are test and reset buttons, and the newer devices have a light-emitting diode (LED) that indicates tripped status. This is a great aid in troubleshooting. GFCIs have feed-through capability. There are line terminals and load terminals. Any downstream wiring or devices connected to the load terminals will

FIGURE 5-2 • A single receptacle is used for dedicated applications where a GFCI is not required, such as a freezer in a below-grade basement.

FIGURE 5-3 • A receptacle is required within 3 ft of a bathroom sink. Of course, it has to be a GFCI.

be GFCI-protected. With a new device, stickers marked "GFCI-Protected" are included, and they should be affixed to the appropriate receptacles. A GFCI does not require an equipment-grounding connection to work, and in fact, the Code permits replacing ungrounded receptacles with GFCIs when no equipment ground is available. If the GFCI load terminals are connected to the electrical supply, the device will be live but no longer have fault-detection capability.

Use of this expedient would void the product listing and also not be Code-compliant where a GFCI is required.

Part I, "General Provisions," states that ratings for other than individual branch circuits are 15, 20, 30, 40, and 50 A. These are very common circuit ratings, and apprentices quickly learn the THHN copper wire sizes that go with each of them. Circuit sizes in amperes are always determined by the overcurrent device rating, even if a larger size wire is used for any reason, such as to limit voltage drop in a long run.

PROBLEM 5-1

How is the circuit size determined?

SOLUTION

The circuit size is always determined by the overcurrent device rating. From the circuit size, the minimum conductor size is derived based on the relevant ampacity table and taking into consideration correction and adjustment factors.

Another topic found in Article 210 is multiwire branch circuits. They are permitted by the Code, with certain limitations, but they are controversial among electricians because, despite certain benefits, they are problematic if not installed correctly.

Article 100, "Definitions," defines a *multiwire branch circuit* as a branch circuit that consists of two or more ungrounded conductors that have a voltage between them and a grounded conductor that has equal voltage between it and each ungrounded conductor of the circuit and that is connected to the neutral or grounded conductor of the system.

The usual arrangement for a multiwire branch circuit is two ungrounded conductors and one grounded conductor connected to a 120/240-V single-phase, three-wire system. Simply put, it is two circuits that share a common neutral. What makes it work is that the two hot wires are connected to opposite legs emanating from the same transformer or power source. Because they are of opposing polarities, the current in the neutral cancels out and is zero when the two circuits are loaded equally. When they are loaded unequally, including when one of them is not loaded at all, the current in the neutral is equal to the difference between the currents in the two legs. For this reason, the current in the neutral will never exceed its overcurrent protection.

There are some advantages to this arrangement. For one thing, it is more economical because three wires do the work of four. This reduces conduit fill, so there are instances when a smaller raceway will suffice. Furthermore, because Type NM cable with two ungrounded conductors is round rather than flat, it is easier to run, and less effort is required to control twisting. It is easier to fish around obstructions. Fewer entries into enclosures are required, and there is less box fill, including in the entrance panel.

Disadvantages

There is a severe downside. To be safe (i.e., not overload the neutral), the two ungrounded conductors have to be fed from opposite legs at the panel. If a mistake is made, you'll burn up the unprotected neutral. In the course of rewiring or repair, an inexperienced worker can inadvertently swing one of the ungrounded conductors to an improper position in the panel so that both hot wires are connected to the same leg. Then, if both circuits become fully loaded, the neutral will carry up to twice the current for which it is protected. The wire can become red hot inside the wall and ignite nearby combustible material. To mitigate this hazard, the current *NEC* edition provides that each multiwire branch circuit is to have a means that will simultaneously disconnect all ungrounded conductors where the branch circuit originates. This will generally consist of a double-pole breaker at the entrance panel. It is always permissible to use two single-pole breakers with a listed handle tie. Trunk slammers are quick to point out that a short length of copper wire inserted through the holes in the single-pole breaker handles will make an improvised handle tie, but this expedient is likely to bend or break, leaving part of the multiwire branch circuit energized.

The Code further seeks to ensure the safety of multiwire branch circuits by providing that the ungrounded and grounded circuit conductors of each multiwire branch circuit are to be grouped by cable ties or similar means in at least one location within the panelboard or other point of origination (Figure 5-4).

FIGURE 5-4 • Cable ties for grouping a multiwire branch circuit.

An exception states that the requirement for grouping does not apply if the circuit enters from a cable or raceway unique to the circuit that makes the grouping obvious.

Still Struggling

Multiwire branch circuits will appear to work if the two ungrounded (hot) conductors are connected to the same phase at the entrance panel or load center, but they are not safe because under some loading conditions the neutral will overheat. Many electricians do not use multiwire branch circuits for this reason. They are prohibited by the *NEC* in some health-care facility and hazardous-location applications.

Article 210 further provides that ungrounded conductors supplied from more than one nominal voltage system are to be identified by phase or line and system at all termination, connection, and splice points. The means of identification may be color coding, marking tape, tagging, or other approved means. [When the Code uses the word *approved*, it means approved by the *authority having jurisdiction* (AHJ), usually the electrical inspector.]

It is further required that the identification means is to be documented in a manner that is readily available or permanently posted at the panel.

Though not specifically required by the *NEC*, electrical trade practice is to identify three-phase circuits as follows:

- 208Y/120-V: black, red, blue
- 480Y/277-V: brown, orange, yellow

Section 210.8, "Ground-Fault Circuit-Interrupter Protection for Personnel," specifies the locations where these lifesaving devices are required, both for dwellings and for nondwellings. You can be sure the exam will have questions on this topic. Most of them will be easy to answer. All you have to do is think back about jobs you have worked on. (The AHJ takes a hard look at this area.) Some questions may be subtly worded and require referring to the Code book. Remember that GFCIs are covered in this article, not in Article 110, "Requirements for Electrical Installations," not in Chapter 3, "Wiring Methods and Materials," and not in Article 406, which covers receptacles in "Equipment for General Use." It is much easier to memorize Code locations than to memorize the many requirements therein.

Section 210.12, "Arc-Fault Circuit-Interrupter Protection," specifies the locations where these highly effective devices are required. Recent code cycles have greatly expanded the number of locations. To find answers to exam questions, just remember that the section on arc-fault circuit interrupters (AFCIs) follows shortly after the section on GFCIs in Article 210.

Part II, "Branch-Circuit Ratings," here again contains material that is sure to be on the exam and that must be observed on the job in order to create a compliant installation. It begins with the very fundamental mandate that branch-circuit conductors are to have an ampacity not less than the maximum load to be served.

Where a load that is continuous (defined in Article 100 as a load where the maximum current is expected to continue for three hours or more) is connected to a branch circuit, the minimum conductor size, before adjustment or correction factors, is to have an allowable ampacity not less than the noncontinuous load plus 125 percent of the continuous load.

Most people who have some knowledge of electrical work assume that the overcurrent device must always be the weakest link in the circuit. This is not always true. Table 210.21(B)(3), "Receptacle Ratings for Various Size Circuits," shows a 15-A receptacle on a 20-A circuit. This is a very common setup in residential and commercial wiring. Another useful table, 210.24, "Summary of Branch-Circuit Requirements," shows size and ratings of various conductors and devices for 15-, 20-, 30-, 40-, and 50-A circuits.

Part III, "Required Outlets," contains the familiar rule that in dwellings, receptacles are to be installed such that no point measured horizontally along the floor line of any wall space is more than six feet from a receptacle outlet. Anyone who has done residential wiring is familiar with this requirement. However, considerable interpretation (e.g., wall dividers, countertop spacing, etc.) arises, and Part III lays it out. Some exam questions may involve subtle distinctions. Remember, all this information is in Article 210, "Branch Circuits."

Article 215, "Feeders," is much shorter. The definition of a *feeder* (Article 100, "Definitions") is all circuit conductors between the service equipment, the source of a separately derived system or other power supply source, and the final branch-circuit overcurrent device. A feeder always has an overcurrent device at each end.

Article 215 contains provisions relating to feeders with a common neutral conductor, feeder equipment-grounding conductor, GFCI protection of personnel, ground-fault protection of equipment, circuits derived from autotransformers, and identification of feeders.

Article 220, "Branch-Circuit, Feeder, and Service Calculations," is perhaps the greatest challenge you will confront in the *NEC*. This article contains provisions that govern sizing out electrical installations, in particular, the conductors at a branch-circuit level and all the way back through the service. Licensing exams have questions that will test your knowledge of and ability to apply Article 220. Because the calculations typically involve more than one step, any small misinterpretation will yield an incorrect answer. At the drawing board or on the job site, a misinterpretation can result in an undersized installation. If it is caught by the inspector, you are facing a massive amount of rework, and if it is not caught by the inspector, the deficient installation could be hazardous. An oversized installation will be costly.

Much of the information used for size calculations is contained in tables. The starting point for many calculations is Table 220.12, "General Lighting Loads by Occupancy." Twenty types of occupancies are listed with their loads in volt-amperes per square foot. It is very instructive to examine this table and consider how loading varies with the type of occupancy. Notice that dwellings, at 3 VA/ft^2, are among the largest energy consumers per area. They are right up there with schools and stores, exceeded only by banks and office buildings at 3.5 VA/ft^2 and greatly exceeding churches and auditoriums, figured at 1 VA/ft^2.

This general lighting load is to be added to the load for receptacle outlets, except for dwellings, where the receptacle load is included in the lighting load. Also added to the lighting load are the loads for specific appliances and other listed items.

The next table that needs to be considered is 220.42, "Lighting Load Demand Factors." This table allows you to reduce the lighting load by a demand factor, expressed as a percentage, based on the type of occupancy and its ampere loading (from Table 220.12). You will notice that the larger the load, the lower (more generous) is the demand factor. This is so because a larger load is less likely to be active at any given time. The phenomenon also varies with the occupancy. For example, dwelling units having a lighting load over 120,000 VA may have a demand factor of 25 percent, whereas the demand factor for the largest load in a warehouse bottoms out at 50 percent. All lighting loads not listed in this table must be figured at 100 percent, that is, no reduction may be taken.

Table 220.54, "Demand Factors for Household Electric Clothes Dryers," allows a 50 percent demand factor for a mere 10 dryers because they have been found to have less likelihood of running simultaneously. Dryers in a commercial laundry have no such allowable reduction because there will be occasions when every dryer is in use.

Table 220.55 applies to household cooking appliances, so it is very commonly used in residential work. Lots of electricians are intimidated when first encountering this table because of its strange form. Most tables that we come across have all entries in the same units. This table is a sort of mixed-media affair. Columns A and B contain demand factors expressed as percentages, like the earlier tables in this article. Column C lists maximum demand in kilowatts, and it is not applicable to appliances with a rating over 12 kW. Additionally, there is a whole page of notes that are applicable to the entire table.

Licensing exams invariably have questions on these demand-factor tables, particularly Table 220.55. It is essential that when you come across a question on household cooking appliances, you know where to find this table and how to apply it.

Table 220.56, "Demand Factors for Kitchen Equipment—Other than Dwelling Units," is applicable to restaurants, hotels, and institutional kitchens. The demand factors are less generous and do not go lower than 65 percent, reflecting the fact that when these kitchens get busy, a greater portion of them will be simultaneously loaded. Note that this loading takes into account not only the number of machines in use but also how fully loaded each of them is.

Part IV, "Optional Feeder and Service Load Calculations," contains an alternate method for sizing out the electrical installation. Some electrical designers routinely figure their jobs on the basis of both standard and optional calculations and then use the smaller size. The optional method apples to a single dwelling only with a 120/240- or 208Y/120-V three-wire service having an ampacity of 100 A or greater. Feeder and service loads for multifamily dwellings, schools, and new restaurants also may be calculated using this optional method. Exams tend to skip over this part, focusing on the standard calculation. If you are confronted with a question on this part, you will know where to find the answer. Provisions are straightforward.

Home and Barn

Part V, "Farm Load Calculations," exists because of the unique properties of electrical usage on a farm. Typically, the farm load is made up of barn and maintenance building usage during the day and heavy consumption in the home in the evening, so these loads may be to some extent noncoincidental.

However, if the farm home has electric heat and the farm has electric grain-drying facilities, Part IV of this article is not to be used to calculate the dwelling load.

Article 225, "Outside Branch Circuits and Feeders," covers outdoor wiring, and a lot of it has to do with clearances and lengths of overhead spans. There is also discussion of required disconnects. Where more than one supply serves a building or facility, the disconnecting means for each supply is to consist of not more than six switches or circuit breakers mounted in a single enclosure, in a group of separate enclosures, or in or on a switchboard. There are to be no more than six disconnects per supply. They are to be grouped in a single location, and each disconnect must be marked to indicate the load served. The service components are upstream of and not protected by the overcurrent devices in the entrance panel. Because they are not protected against overcurrent by utility fusing except at a much higher level than their ampacity, to compensate, they are built to a higher standard. Service cable may be either installed in raceway or in concentric construction (the neutral being a grounded outer braid that shields the hot conductors), so there is greater protection against physical damage. Often the service conductors are underground or on the outside of the building, making them less of a fire hazard.

Part I, "General," begins with the provision that a building or structure is to be supplied by a single service only. This is a very fundamental principle in electrical wiring because more than one service in a single building could give rise to a situation where users or maintenance workers would assume that the power was disconnected when in fact portions of the premises wiring could be energized. Furthermore, adjacent wiring and equipment could be energized from separate utility transformers, leading to dangerous differences in voltage potentials.

Nevertheless, some exceptions permit multiple services at a single building or structure under limited circumstances:

- *Special conditions*. Additional services may supply fire pumps, emergency systems, legally required standby systems, optional standby systems, parallel power-production systems, and systems designed for connection to multiple sources of supply for the purpose of enhanced reliability.
- *Special occupancies.* By special permission, additional services are permitted for multiple-occupancy buildings where there is no available space for service equipment accessible to all occupants and where a single building or structure is sufficiently large to make two or more services necessary.

- *Capacity requirements.* Additional services are permitted where the capacity requirements are in excess of 2,000 A at a supply voltage of 600 V or less, where the load requirements of a single-phase installation are greater than the serving agency normally supplies through one service, or by special permission.
- *Different characteristics.* Additional services are permitted for different voltages, frequencies, or phases or for different usages such as different rate schedules.

Conductors other than service conductors are not to be installed in the same service raceway or service cable.

TIP ***A frequent Code violation involves using a service mast to secure telephone or cable television (CATV) conductors. The purpose of the service mast is to provide the required ground clearance for the service drop.***

Where a service raceway enters a building or structure from an underground distribution system, it is to be sealed. Clearances for overhead and underground service conductors are specified. Driving around a suburban area, you will see many service configurations, and this is a good way to get a sense of the possible ways to solve clearance challenges.

Article 240, "Overcurrent Protection," provides that a fuse or an overcurrent trip unit of a circuit breaker is to be connected in series with each ungrounded conductor. A combination of a current transformer and an overcurrent relay is equivalent to an overcurrent trip unit. It is further stated (with exceptions) that circuit breakers are to open all ungrounded conductors of the circuit both manually and automatically.

Part II, "Location," states the principle that overcurrent protection is to be provided for each ungrounded circuit conductor and is to be located at the point where the conductors receive their supply. This is the usual scenario for fuses and circuit breakers. There are, however, eight exceptions to this general principle, and they comprise the infamous tap rules. These allow, under strictly limited conditions, conductors to be protected at more than their ampacity or at the downstream end. These tap conductors are limited in length and must be physically protected, for example, by being enclosed in a raceway.

By taking advantage of these tap rules, electricians are able to save on large amounts of materials and labor in certain installations, but the rules must be carefully observed to avoid creating a noncompliant or hazardous installation. The principle tap rules are

(A) Branch-circuit tap conductors:

- Conductors tapped from a 50-A branch circuit supplying electric ranges, wall-mounted electric ovens, and counter-mounted electric cooking units are to have an ampacity of not less than 20 A and are to be sufficient for the load to be served. The taps must not be longer than necessary for servicing the appliance.
- The neutral conductor of a three-wire branch circuit supplying a household electric range, a wall-mounted oven, or a counter-mounted cooking unit is permitted to be smaller than the ungrounded conductors where the maximum demand of a range of 8¾-kHz or more rating has been calculated according to column C of Table 220.55, but such conductors may not have an ampacity of less than 70 percent of the branch-circuit rating and are not to be smaller than 10 AWG.
- For other loads, tap conductors are to have an ampacity sufficient for the load served. In addition, they are to have an ampacity of not less than 15 A for circuits rated less than 40 A and not less than 20 A for circuits rated at 40 or 50 A and only where these tap conductors supply any of the following loads:
  (a) Individual lampholders or luminaires with taps extending not longer than 18 in. beyond any portion of the lampholder or luminaire
  (b) A luminaire having tap conductors of a type suitable for the temperature encountered running from the luminaire terminal connection to an outlet box placed at least 1 ft from the luminaire (These conductors are to be in suitable raceway or Type AC or MC cable of at least 18 in. but not more than 6 ft in length.)
  (c) Individual outlets, other than receptacle outlets, with taps not over 18 in. long
  (d) Infrared lamp industrial heating appliances
  (e) Nonheating leads of deicing and snow-melting cables and mats

Tap rules for feeders are as follows: Feeder conductors are permitted to be tapped, without overcurrent protection at the tap, where

(1) Taps are not over 10 ft long if the length of the conductors does not exceed 10 ft and the tap conductors comply with all the following:
- The ampacity of the tap conductors is not less than the combined calculated load on the circuits supplied by the tap conductors and not less than the rating of the device supplied by the tap conductors or not less than the rating of the overcurrent protective device at the termination of the tap conductors.

- The tap conductors do not extend beyond the switchboard, panelboard, disconnecting means, or control devices they supply.
- Except at the point of connection to the feeder, the tap conductors are enclosed in a raceway that is to extend from the tap to the enclosure of an enclosed switchboard, panelboard, or control device or to the back of an open switchboard.
- For field installations, if the tap conductors leave the enclosure or vault in which the tap is made, the ampacity of the tap conductors is not less than one-tenth the rating of the overcurrent device protecting the feeder conductors.

(2) Taps not over 25 ft long: Where the length of the tap conductors does not exceed 25 ft and the tap conductors comply with all the following:

- The ampacity of the tap conductors is not less than one-third the rating of the overcurrent device protecting the feeder conductors
- The tap conductors terminate in a single circuit breaker or a single set of fuses that limit the load to the ampacity of the tap conductors (This device is permitted to supply any number of additional overcurrent devices on its load side.)
- The tap conductors are protected from physical damage by being enclosed in an approved raceway or by other approved means.

(3) Taps supplying a transformer: primary plus secondary not over 25 ft long: Where the tap conductors supply a transformer and comply with all the following conditions:

- The conductors supplying the primary of a transformer are to have an ampacity that is at least one-third the rating of the overcurrent device protecting the feeder conductors.
- The conductors supplied by the secondary of the transformers are to have an ampacity that is not less than the value of the primary-to-secondary voltage ratio multiplied by one-third of the rating of the overcurrent device protecting the feeder conductors.
- The total length of one primary plus one secondary conductor, excluding any portion of the primary conductor that is protected at its ampacity, is not over 25 ft.
- The primary and secondary conductors are protected from physical damage by being enclosed in an approved raceway or by other approved means.
- The secondary conductors terminate in a single circuit breaker or a set of fuses that limit the load current to not more than the conductor ampacity that is permitted in the Chapter 3 ampacity tables.

(4) Taps over 25 ft long: Where the feeder is in a high-bay manufacturing building over 35 ft high at walls and the installation complies with all the following conditions:

- Conditions of maintenance and supervision ensure that only qualified persons service the systems.
- The tap conductors are not over 25 ft long horizontally and not over 100 ft total length.
- The ampacity of the tap conductors is not less than one-third the rating of the overcurrent device protecting the feeder conductors.
- The tap conductors terminate at a single breaker or a single set of fuses that limit the load to the ampacity of the tap conductors. This single overcurrent device is permitted to supply any number of additional overcurrent devices on its load side.
- The tap conductors are protected from physical damage by being enclosed in an approved raceway or by other approved means.
- The tap conductors are continuous from end-to-end and contain no splices.
- The tap conductors are sized 6 AWG copper or 4 AWG aluminum or larger.
- The tap conductors do not penetrate walls, floors, or ceilings.
- The tap is made no less than 30 ft from the floor

(5) Outside taps of unlimited length: When the conductors are located out doors of a building or structure, except at the point of load termination, and comply with all the following conditions:

- The conductors are protected from physical damage in an approved manner.
- The conductors terminate at a single circuit breaker or a single set of fuses that limit the load to the ampacity of the conductors. This single overcurrent device is permitted to supply any number of additional overcurent devices on its load side.
- The overcurrent device for the conductors is an integral part of a disconnecting means or is to be located immediately adjacent thereto.
- The disconnecting means for the conductors is installed at a readily accessible location outside a building or structure, inside, nearest the point of entrance of the conductors, or where installed in accordance with Section 230.6, nearest the point of entrance of the conductors. (Section 230.6 states criteria for ascertaining whether conductors are outside of a building. For example, conductors installed within a building that are encased in brick or concrete not less than 2 in. thick are considered "outside" that building.)

PROBLEM 5-2

Why are outside taps allowed (with conditions) to be of unlimited length?

SOLUTION

Outdoor locations are less hazardous in terms of electrical fire because there is less chance that combustible material will ignite.

(C) Transformer secondary conductors: A set of conductors feeding a single load or each set of conductors feeding separate loads is permitted to be connected to a transformer secondary, without overcurrent protection at the secondary, as follows:

- Protection by primary overcurrent device: Conductors supplied by the secondary side of a single-phase transformer having a two-wire (single-voltage) secondary or a three-phase, delta-delta-connected transformer having a three-wire (single-voltage) secondary is permitted to be protected by overcurrent protection provided on the primary (supply) side of the transformer, provided that this protection does not exceed the value determined by multiplying the secondary conductor ampacity by the secondary-to-primary transformer voltage ratio.
- Transformer secondary conductors are not over 10 ft long and complying with all the following:
  (1) The ampacity of the secondary conductors is not less than the combined calculated loads on the circuits supplied by the secondary conductors and is not less than the rating of the device supplied by the secondary conductors or not less than the rating of the overcurrent protective device at the termination of the secondary conductors.
  (2) The secondary conductors do not extend beyond the switchboard, panelboard, disconnecting means, or control devices they supply.
  (3) The secondary conductors are enclosed in a raceway that extends from the transformer to the enclosure of an enclosed switchboard, panelboard, or control device or to the back of an open switchboard.
  (4) For field installations where the secondary conductors leave the enclosure or vault in which the supply connections are made, the rating of the overcurrent device protecting the primary of the transformer, multiplied by the primary-to-secondary transformer voltage ratio, is not to exceed 10 times the ampacity of the secondary conductor.

- Industrial installation of secondary conductors not over 25 ft long: For industrial installations only, where the length of the secondary conductors does not exceed 25 ft, comply with all the following:
  (1) Conditions of maintenance and supervision ensure that only qualified persons service the systems.
  (2) The ampacity of the secondary conductors is not less than the secondary current rating of the transformer, and the sum of the ratings of the overcurrent devices does not exceed the ampacity of the secondary conductors.
  (3) All overcurrent devices are grouped.
  (4) The secondary conductors are protected from physical damage by being enclosed in an approved raceway or by other approved means.
- Outside secondary conductors: Where the conductors are located outdoors of a building or structure, except at the point of load terminations, comply with all the following conditions:
  (1) The conductors are protected from physical damage in an approved manner.
  (2) The conductors terminate at a single circuit breaker or a single set of fuses that limit the load to the ampacity of the conductors. This single overcurrent device is permitted to supply any number of additional overcurrent devices on its load side.
  (3) The overcurrent device for the conductors is an integral part of a disconnecting means or is to be located immediately adjacent.
  (4) The disconnecting means for the conductors is installed at a readily accessible location complying with one of the following: Outside of a building or structure or inside, nearest the point of entrance of the conductors.
- Secondary conductors from a feeder-tapped transformer: Transformer secondary conductors installed in accordance with the provisions for industrial installations (noted earlier) are permitted to have overcurrent protection as specified in that section.
- Secondary conductors not over 25 ft long, complying with all the following:
  (1) The secondary conductors are to have an ampacity that is not less than the value of the primary-to-secondary voltage ratio multiplied by one-third the rating of the overcurrent device protecting the primary of the transformer.

(2) The secondary conductors terminate in a single circuit breaker or set of fuses that limit the load current to not more than the conductor ampacity as permitted by the ampacity tables in Chapter 3.

(3) The secondary conductors are protected from physical damage by being enclosed in an approved raceway or by other approved means.

(D) Service conductors are permitted to be protected by overcurrent devices in accordance with Article 230, "Services."

(E) Busway taps are covered in Article 368, "Busways."

(F) Motor-circuit taps are covered in Article 430, "Motors, Motor Circuits, and Controllers."

(G) Conductors from generator terminals are covered in Article 445, "Generators."

(H) Battery conductors are to have overcurrent protection installed as close as practicable to the battery terminals.

The foregoing survey of tap rules, together with other instances where overcurrent protection is not installed at the upstream end of conductors, illustrates the fact that when standard overcurrent protection is not put in place at the upstream end of the entire conductor, other compensatory types of protection are mandated. Furthermore, strict limits are placed on the installation, such as length of the tap conductor and the settings in which it is permitted.

All the details are too extensive for most of us to have in mind at all times, but what we can do is look up the requirements in the *NEC* as needed to answer exam questions or size out an installation. You just need to know where to look. Most of the tap rules are found in Part II, "Location," of Article 240, "Overcurrent Protection," and therein are also found relevant references to other parts of the Code.

Additional subtopics in Article 240, less frequently referenced but sometimes appearing in licensing exams, are Part III, "Enclosures," Part IV, "Disconnecting and Guarding," Part V, "Plug Fuses, Fuseholders and Adapters," Part VI, "Cartridge Fuses and Fuseholders," and Part VII, "Circuit Breakers."

Grounding versus Bonding

Article 250, "Grounding and Bonding," is lengthy. The topic is complex, but the article is well organized, so the whole subject, as covered in the *NEC*, lends itself to study and, if not total mastery, at least familiarity to a point where problems that arise in an exam or during an installation can be quickly resolved. A principal object of correct bonding and grounding is electrical safety, and to this end, a whole protocol of conductors and connections has been developed, and it's all in Article 250.

Many electricians neglect the importance of bonding in favor of grounding. Whereas a good connection to one or more ground rods or other grounding electrodes is essential for all electrical installations, this is not the whole story. We'll consider grounding and bonding as two separate topics—how they are accomplished and the separate purposes they serve.

To begin, it must be conceded that there is such a thing as an ungrounded system, and in some very limited settings, it offers distinct benefits. But it should be stressed than even an ungrounded system requires an equipment-grounding conductor to keep exposed conductive, normally non-current-carrying surfaces at ground potential so that they do not present a shock hazard.

Electrical systems that have a nominal voltage under 50 V are generally not required to be grounded unless they are supplied by a transformer whose supply to the primary exceeds 150 V to ground, where supplied by transformers if the transformer supply system is ungrounded, or where installed outside as overhead conductors. We see many low-voltage systems such as power-limited fire alarm conductors and thermostat circuits that control steam heat or residential furnaces in which both conductors are isolated from ground.

Much less common are these alternating-current (ac) systems of 50 V to less than 1,000 V that are not required to be grounded:

- Electrical systems used exclusively to supply industrial electric furnaces for melting, refining, tempering, and the like.
- Separately derived systems used exclusively for rectifiers that supply only adjustable-speed industrial drives.
- Separately derived systems supplied by transformers that have a primary voltage rating of less than 100 V, provided that all the following conditions are met:
  (a) The system is used exclusively for control circuits.
  (b) The conditions of maintenance and supervision ensure that only qualified persons service the installation.
  (c) Continuity of control power is required.

The classic example of a system of this sort is the control circuit for a crane-mounted electromagnet that is used to lift and move scrap iron in an industrial setting. If there is a ground fault, the control circuit will no longer maintain power to the magnet, allowing the heavy load to drop. Hazards of this nature are mitigated by using an ungrounded system, at the cost of sacrificing automatic ground-fault protection. To compensate, audiovisual alarms are built into the system to alert workers so that an orderly shutdown can commence, and then repairs are completed before putting the system back in service.

To emphasize, when a system is designated ungrounded, it means that the current-carrying conductors are all isolated from ground, but an equipment-grounding conductor is always required. Ungrounded systems of 50 to 1,000 V are not seen very often. Most systems of 50 to 1,000 V have a grounded current-carrying neutral conductor and an also equipment-grounding conductor that is connected to metal raceways, enclosures, and non-current-carrying conductive structures that could become energized. If this happens, the equipment-grounding conductor will carry the fault current back to the service-entrance panel, allowing sufficient current flow in the ungrounded (hot) conductor so as to trip the overcurrent device and shut down the faulted circuit. Accordingly, the grounded and grounding conductors are usually at very close to the same potential. They are electrically connected at the service equipment by means of the main bonding jumper. They should not be connected anywhere else downstream. That would be a Code violation because it makes for harmful circulating currents. In many electrical matters, including grounding and bonding, redundancy is a good thing, but this is not the case in regard to the connection between a grounded conductor and a grounding conductor. They should be connected at the service equipment only, by means of the main bonding jumper.

The equipment-grounding conductor is not always a wire. If you are doing residential wiring using Romex (Type NM) cable, the equipment-grounding conductor is a bare copper wire that is included within the same jacket as the insulated current-carrying conductor. If you are installing a raceway system, options may be available as far as the equipment-grounding conductor is concerned. You can install a green insulated equipment-grounding conductor, and this is always a good choice. Alternately, the equipment-grounding conductor is not always a wire. Section 250.118 lists the types of conductors that qualify as equipment-grounding conductors:

- A copper, aluminum, or copper-clad aluminum conductor (It may be solid or stranded, insulated, covered, or bare and in the form of a wire or a busbar of any shape.)
- Rigid metal conduit
- Intermediate metal conduit

- Electrical metallic tubing
- Listed flexible metal conduit meeting all the following conditions:
  (a) The conduit is terminated in listed fittings.
  (b) The circuit conductors contained in the conduit are protected by overcurrent devices rated at 20 A or less.
  (c) The combined length of flexible metal conduit and flexible metallic tubing and liquid-tight flexible metal conduit in the same ground-fault current path does not exceed 6 ft.
  (d) If used to connect equipment where flexibility is necessary to minimize the transmission of vibration from the equipment or to provide flexibility for equipment that requires movement after installation, an equipment-grounding conductor is to be installed.
- Liquid-tight flexible metal conduit, flexible metallic tubing, armor of Type AC cable, the copper sheath of mineral-insulated, metal-sheathed cable, Type MC cable, cable trays, cable bus framework, other listed electrically continuous metal raceways, and listed auxiliary gutters and surface metal raceways listed for grounding. Some of the preceding have conditions attached.

A very common setup is electrical metallic tubing (EMT), where redundant grounding is not required, except in some sensitive locations, such as in a patient-care area of a health-care facility (Figure 5-5). Notice that the flexible conduits are restricted in length if they are to be used as equipment-grounding conductors. This is so because if longer, they are more likely to lose low-impedance connectivity. Of course, nonmetallic flexible raceway, which has a similar outer appearance, can never be used as an equipment-grounding conductor.

FIGURE 5-5 • EMT couplings: dry location on left, wet location on right.

It is always better to install a green wire if conduit fill will allow. Many electricians "pull a green for everything" and bond all enclosures with a wire equipment-grounding conductor. You will undoubtedly come across instances where a length of EMT has separated from a coupling or connector, and here a green wire could quite literally be a lifesaver.

TIP ***If ground continuity is lost and there is a fault to ground, the overcurrent device will not trip, and non-current-carrying conductive services may remain live.***

I have mentioned that grounding and bonding are two separate, distinct operations, though closely related. Bonding, in its simplest form, consists of connecting all normally non-current-carrying conductive elements that are likely, in the event of a fault, to become energized to an equipment-grounding conductor. This is run with the current-carrying conductors within the raceway (or it is the raceway itself), or it is run with the current-carrying conductors in a cable. It goes back to the entrance panel, where it is terminated at the grounding terminal bar and, through the main bonding jumper, to the neutral bar. The current path must be of sufficiently low impedance to keep the non-current-carrying conductive elements at ground potential and facilitate operation of the overcurrent device.

On the subject of bonding, there are some additional Code requirements that are sometimes neglected. For one thing, any interior metal water pipe is to be bonded to the grounding means at the service equipment. This is done by running a 6 AWG copper conductor, solid or stranded, bare, covered, or insulated, from a water pipe, where it is connected by means of a grounding clamp, back to the grounding terminal at the service-entrance panel. Be sure that the entire metallic water system is continuous or that there are bonding jumpers around the nonmetallic portions.

Another bonding task is to ensure that communications and data technicians bond their equipment to the electrical grounding system. Examples are telephone, CATV, television, and Internet satellite dish systems and broadband service. So that these technicians can make their grounding connections, the premises wiring electrician supplies what is known as an *intersystem bonding termination*. This is an inexpensive piece of hardware that mounts on the exterior siding. It has a protective cover that can be removed to access a terminal strip with provision for multiple grounding connections. The electrician wires the intersystem bonding termination to a grounding lug in the meter socket enclosure. Even lightning-protection systems are to be bonded to all other grounding systems in this fashion. Home owners sometimes have a problem with this, fearing the lightning will be drawn into their home, but this bonding is essential from the standpoint of safety to ensure that dangerous differences in electrical potential do not exist between these different grounding systems within or on the same building.

As for grounding, that is a different matter. Let's start at the earth. This connection is made by means of the grounding electrode(s). An example is the familiar ground rod. (Usually a single ground rod is not sufficient.) Besides a ground rod, several other types of grounding electrodes are used. An underground metal waterline, especially if connected to a metal well casing or water main, makes an excellent grounding electrode, but there are problems that may arise. These days most waterline work is done in plastic or, more properly, PVC. The indoor section, up to the point where it penetrates the basement wall, may be copper or galvanized steel, but the outdoor portion may have been redone in PVC, or a segment may have been replaced in the course of a repair, breaking electrical continuity.

Besides metal waterlines, another permitted grounding electrode is the metal frame of a building, provided that either

- It is connected to at least one structural member that is in direct contact with the earth for 10 ft or more, with or without concrete encasement, or
- It is connected to hold-down bolts securing the structural steel column that are connected to a concrete-encased electrode described below, located in the support footing or foundation. The hold-down bolts are to be connected to the concrete-encased electrode by welding or steel wires.

A very effective grounding method is the concrete-encased electrode, but it requires advance planning, before the concrete foundation is poured. It is to consist of at least 20 ft of either of the following:

- One or more bare or zinc-galvanized or other electrically conductive coated-steel reinforcing bars or rods not less than ½ in. in diameter, installed in one continuous 20-ft length, or if in multiple pieces connected together by the usual steel tie wires, welding, or other effective means to create a 20-ft or greater length, or
- Bare copper conductor not smaller than 4 AWG.

Metallic components must be encased by at least 2 in. of concrete and are to be located horizontally within that portion of a concrete foundation or footing that is in direct contact with the earth or within vertical foundations or structural components or members that are in direct contact with the earth. If the concrete is placed over foam or plastic, as is sometimes the practice, this type of grounding electrode will not work. Also, PVC-encased rebar, sometimes used to prevent corrosion, will negate effective grounding. Otherwise, this is probably the best grounding system.

A ground ring encircling the building below grade, consisting of at least 20 ft of copper conductor not smaller than 2 AWG is another good option.

Plate electrodes must expose at least 2 ft^2 of surface to the soil. If made of iron or steel, they are to be at least ¼ in. thick. Nonferrous plates, usually copper, are to be 0.06 in. thick.

Other local underground piping systems, tanks, and so forth may be used as well. Metal underground gas piping systems are not to be used as grounding electrodes. However, they should be bonded to the electrical grounding system. They just can't be counted as the grounding electrode. A gas furnace or cookstove usually will have a 120-V ac power supply, which will include an equipment-grounding conductor. This will provide grounding and bonding for the gas line.

I have referred to different grounding electrodes as being effective in various degrees. This refers to the extent to which they will provide a low-impedance connection to the earth. The connection is measured in ohms, but an ordinary ohmmeter cannot take this measurement. At the very least, a zero-impedance ground electrode would have to be established, and the whole purpose of the exercise would be defeated.

Measuring Ground Resistance

There is an instrument that will measure ground resistance, and it can be used to check the quality of a ground electrode connection. This meter will cost between $500 and $5,000. It is not used for most ordinary installations. The Code provides that a single rod, pipe, or plate electrode is to be supplemented with an additional grounding electrode. An exception states that if a single electrode has a resistance to earth of 25 Ω or less, the supplemental electrode is not required. Rather than performing the test, most electricians install a second ground rod. If this is done, no further ground-resistance measurement is required.

If multiple rods, pipes, or plates are installed, they must be at least 6 ft apart. If they are closer, they are less effective. Although not a Code requirement, grounding electrodes should be situated along the drip line from the roof because soil moisture makes for less ground resistance. In dry, gravelly soils, achieving good grounding is more of a problem. There are additives that can be mixed with the soil surrounding a grounding electrode, but they are rarely used.

A ground rod should be driven so that the top end is below grade, and the ground-rod clamp, listed for burial, is out of sight. This discourages vandalism and also contributes to ground conductivity. If bedrock is encountered, the top of the rod should never be cut off. Instead, you may drive it at a maximum 45-degree angle or make a trench and lay it horizontally with at least 30 in. of ground cover. In difficult cases, you can bring in fill material or go to a different type of grounding-electrode system.

Auxiliary grounding electrodes can be placed anywhere along a branch circuit if the need is felt for extra protection. However, there should never be a *floating ground*, which is a portion of a branch circuit including the load that is connected to a grounding electrode but has no equipment-grounding conductor that goes back to the electrical supply. Manufacturers of outdoor lighting with metal towers for parking lots sometimes specify that an extra ground rod is to be provided at each unit. Because the manufacturer's instructions are a part of the product listing, it is necessary to provide this grounding even though it is not required by the *NEC*.

The grounding electrode conductor connects to the ground rod by means of a listed ground-rod clamp. (Trunk slammers, always resourceful, use hose clamps or nut and bolt arrangements. These expedients will corrode in time or burn out if there is a voltage spike caused by lightning or by a grounded utility line.)

The grounding-electrode conductor may be solid or stranded, bare, covered, or insulated. This wire must be run in a raceway if it is smaller than 6 AWG. Otherwise, it may be run along a building surface. A grounding-electrode conductor is to be installed in one continuous length without joint or splice. It is run to the service-entrance enclosure where the main disconnect is located or to the meter-socket enclosure where a lug is provided for making this connection. The conductor is sized out in accordance with Table 250.66 based on the size of the service-entrance conductors supplying the individual service disconnecting means. However, it never needs to be larger than 6 AWG copper where rod, pipe, or plate electrodes are used, 4 AWG copper for connections to concrete-encased electrodes, or for a ground ring and no larger than the conductor used for the ground ring.

TIP ***There are a number of ground-electrode options. The familiar ground rod is used most commonly but may not be best in sensitive applications.***

Article 250, as I have mentioned, is one of the most fundamental of Code articles, and you may expect it to be the subject of exam questions. There are not a lot of calculations, as in Article 220, but plenty of individually focused requirements to be looked up. If you know where to find them, the answers are straightforward.

Chapter 2 concludes with two articles, 280, which covers surge arrestors over 1 kV, and 285, which covers surge-protective devices of 1 kV or less. Formerly, these devices were distinguished by where they were located, line side or load side of the main disconnect. The current Code delineates them on the basis of voltage. Keeping this in mind, you can go to the correct article to find the answers you need.

This takes care of *NEC* Chapter 2. It is a good idea to go through it periodically because it is the heart of the Code and crucial for all electrical work.

Chapter 3 is the final part of the *NEC* that applies generally, although it will be found to be somewhat more specific than the preceding chapters. It is titled "Wiring Methods and Materials." The articles in Chapter 3, while broad in scope, are full of specific information. If you pay close attention to the contents and structure of Chapter 3, you will find answers as needed. We will begin at the beginning, Article 300, "Wiring Methods," and then we will survey the rest of the chapter.

Article 300, following the usual Code template, begins with a statement of scope. It notes that the provisions of the article do not apply to conductors that are an integral part of equipment, such as motors, controllers, motor control centers, factory-assembled control equipment, or listed utilization equipment.

Section 300.3 specifies that single conductors are to be installed only where part of a recognized wiring method, primarily in cables or raceways. There are a few exceptions, such as overhead conductors, grounding-electrode conductors, and wiring within photovoltaic (PV) solar arrays, but the basic rule is that you are not supposed to have individual conductors secured to a building surface or concealed behind finish surfaces, much less flying through the air. Violations are seen when submersible-pump cable is brought into a building to terminate at the control box. When the red, black, and yellow conductors are stapled to the framing, fly through the air, and are pinched in a Romex connector, it's not the right way to do it. It takes a little more work to run the pump cable in a raceway, but it makes a better installation.

All conductors of the same circuit, including the grounded conductor and all equipment-grounding conductors and bonding conductors, are to be contained within the same raceway, cable, or other wiring method. There are exceptions, and you can refer to them, but this is the basic rule. When violated, there can be inductive heating. Also, tracing wiring for the purpose of maintenance could become chaotic.

It is provided that conductors of alternating-current (ac) and direct-current (dc) circuits rated 600 V nominal or less are permitted to occupy the same wiring enclosure, cable, or raceway. All conductors are to have an insulation rating equal to at least the maximum circuit voltage applied to any conductor within the enclosure, cable, or raceway.

Conversely, conductors of circuits rated over 600 V nominal are not to occupy the same enclosure, cable, or raceway with conductors of circuits rated 600 V or less. There are exceptions, such as the secondary wiring of electric-discharge lamps, but the basic rule is to keep them separate. As is often the case with Code provisions, 600 V is the cutoff. Notice that the commonly used 600-V nominal system falls into the lower category.

PROBLEM 5-3

Why is it necessary for all conductors in the same enclosure, raceway, or cable to be insulated for the highest voltage present?

SOLUTION

If a higher-voltage conductor faulted to a lower-voltage conductor, the greater amount of electrical energy could present a shock or fire hazard.

Section 300.4, "Protection against Physical Damage," contains provisions relating to how wiring is to be installed so as not to be subject to damage from stray nails and screws or chafing where entering enclosures or damage at structural joints in a building where relative motion can take place—expansion, contraction, or deflection.

Minimum Cover

Section 300.5 concerns underground installations. Minimum cover requirements are given in Table 300.5. (A similar table, 300.50, is located later in the article for installations over 600 V.) The table is easy to read—various minimum cover requirements for different locations and wiring methods. If you get into a problem with bedrock, for example, you can use rigid conduit and go as shallow as 6 in. in some locations. If you forget where to find the minimum cover requirements, look in the Index for underground wiring.

Other topics in Article 300 include mechanical and electrical continuity of conductors; length of free conductors at outlets, junctions, and switch points;

and raceway installations. Many workmanship issues are treated. For example, with one exception, all raceway systems are to be completed and connected to the enclosures at both ends before the conductors are installed. Underground conduit should be completed and backfilled before the wire is pulled. However, a pull rope can be inserted as the raceway is assembled, piece by piece.

Article 310, "Conductors for General Wiring," is probably the most frequently referenced material in the entire *NEC*. The tables toward the end of the article give allowable ampacities of different size conductors under varying conditions, and this information allows you to size out first the conductors and then the raceway for any electrical installation.

Before these tables, there are some sections that cover situations encountered in the course of many electrical installations. One of the most important of them is Section 310.10(H), "Conductors in Parallel." It is permitted, and considered good practice, to run two or more conductors in parallel instead of a single large one in high-current applications where a single conductor would be difficult to bend and terminate owing to its size. Aluminum, copper-clad aluminum, or copper conductors for each phase or polarity and neutral or grounded conductors are permitted to be connected in parallel (electrically joined at both ends) only in sizes 1/0 AWG and larger if in compliance with the following:

- Conductor characteristics: The paralleled conductors in each phase, polarity, neutral, grounded circuit conductor, equipment-grounding conductor, or equipment-bonding jumper are to be the same length, consist of the same conductor material, be the same size in circular mil area, have the same insulation type, and be terminated in the same manner.
- Where run in separate cables or raceways, the cables or raceways are to have the same number of conductors and the same electrical characteristics.
- Conductors installed in parallel must comply with Section 310.15(B)(3)(a), which provides for adjustment factors for more than three current-carrying conductors in a raceway or cable. This is an important requirement in figuring conductor sizes, and I'll discuss it presently. For now, remember that as with all wiring, these factors are applicable in parallel installations.
- Equipment-grounding conductors and equipment-bonding jumpers are to be sized out based on 250.122 and 250.102. Accordingly, in some instances they will be smaller than 1/0 AWG.

Tables 310.15(B)16 through 310.15(B)21 give ampacities under various conditions for conductors rated 0 to 2,000 V. (A later set of tables gives ampacities for conductors rated 2,001 to 35,000 V.)

The usual procedure is to begin with the desired ampacity based on the calculated load, then work back to the conductor size, and from there choose the raceway size complying with conduit fill requirements.

Before consulting the relevant ampacity table, though, it is necessary to apply two factors to find the final ampacity that you are going to use. The order in which the factors are applied does not matter. You'll get the same answer either way as long as both factors, where needed, are applied.

The whole thing comes down to heat. When current passes through a wire, heat is generated, and if heat is generated faster than it is dissipated, the temperature will rise. There are several aspects that need to be considered. For one thing, it is damage to the insulation that concerns us, and whether or not this damage takes place depends on the type of insulation. Repeated heating episodes degrade the insulation, increasing the chance of shock, short circuit, and fire.

The ambient temperature is important. If the cable or raceway passes through a boiler room where there is an elevated temperature, the enclosed conductors will not be able to get rid of the heat generated owing to the current they carry unless the ampacity is corrected. Furthermore, if current-carrying conductors are run in close proximity, that is, within the same raceway or cable, their ability to dissipate heat is diminished.

Still Struggling

The problem with heat is that it damages conductor insulation, and the problem is cumulative over time. When insulation becomes brittle or charred, it ignites more easily, that is, at a lower temperature. Moreover, a conductor that is loaded beyond its capacity may become hot enough to ignite nearby combustible material.

Ambient temperature-correction factors based on 30°C (86°F) are given in Table 310.15(B)(2)(A). Notice that ambient temperatures in the Celsius scale are shown in the left column and in the Fahrenheit scale are shown in the right column, so you can use either one. Temperature ratings of conductors, which have to do with the type of insulation, are shown in the three middle columns, for 60, 75, and 90°C. (These temperature ratings are printed on the conductor insulation and are customarily given in Celsius numbers.) A subtitle to the table states that for ambient temperatures other than 30°C, multiply the allowable ampacities specified in the ampacity tables by the appropriate correction factor shown in the table. For lower ambient temperatures, the correction factor is above 1.0, so this means that you will get a higher allowable ampacity, which translates to a smaller conductor size. For ambient temperatures above 26 to 30°C, the correction factor is less than 1.0, so the allowable ampacity is reduced, making for a larger conductor size.

This is all very straightforward, and with the miraculous number-crunching ability of the handheld calculator, these factors are easy to apply. Notice that these numbers are called *correction factors*, whereas the numbers applicable to current-carrying conductor density are known as *adjustment factors*. Despite the difference in terminology, they are both applied to allowable ampacity in the same way.

Table 310.15(B)(3)(a) gives adjustment factors for more than three current-carrying conductors in a raceway or cable. If there are three or less current-carrying conductors in a raceway or cable, no adjustment to the allowable ampacity is required. If there are four to six such conductors, the allowable ampacity has to be multiplied by 80 percent. The percentage decreases as the number of current-carrying conductors increases. If there are 41 or more current-carrying conductors in the raceway or cable, the allowable ampacity is multiplied by 35 percent. In other words, it is only a little over a third of the ampacity where there are three or fewer current-carrying conductors in the same raceway or cable.

These factors are easy to apply once you come to an understanding of what is meant by a current-carrying conductor. For one thing, the conductor count applies only to power and lighting conductors. It does not apply to conductors of communications and data systems. Second, an equipment-grounding conductor is never considered current-carrying. The hot-phase wires are always current-carrying. What about the neutral? In a 240-V balanced load, the two legs are 180 degrees out of phase, so the two currents in the neutral cancel out. The neutral is not current-carrying. When these two loads are not equal, the neutral carries the difference. If there is a gray area or any uncertainty, it is better to count the neutral as current-carrying to make sure that you don't overheat the wires. It is possible to comply with this requirement either by multiplying the allowable ampacity by the adjustment factor or by separating some of the circuits into more than one cable or raceway. In no event, however, should a circuit be split up so that its conductors are routed separately, as pointed out earlier.

Before we get to the ampacity tables, there is one other table that we should look at. It is Table 310.15(B)(7), "Conductor Types and Sizes for 120/240-V, Three-Wire, Single-Phase Dwelling Services and Feeders." This table applies to certain conductor types, including THHN, SE, and USE.

Things to note about this table are that it applies only to dwellings with single-phase services. The table permits you to use smaller conductors for services and feeders than required by the ampacity tables. This is so because of diversity of load, and to take two familiar examples, a 100-A service call for 2 AWG aluminum, and a 200-A service call for 4/0 AWG aluminum. Service sizes run up to 400 A. The table is very simple and easy to use, but it may be hard to find because it is located in an obscure corner of Chapter 3.

This brings us to the main ampacity tables. The first of them, 310.15(B)(16), is by far the most frequently referenced, and it will most likely be the one you will need. It applies to conductors in cables, raceways, and direct-burial installations with temperature ratings of 60, 75, and 90°C ratings, up to and including 2,000 V. The other tables are for less common installations, such as higher temperature and voltage ratings, conductors in free air rather than raceways and cable, and so on. Most exam questions take you to the first table, but read the wording carefully to be sure.

Notice that this table has two parts, copper conductors on the left half of the table and aluminum conductors on the right. Thus, for most exam questions, it is going to be the left side that concerns us. The outside columns are wire sizes, repeated on both sides of the table for ease of use.

Prior to the exam, you should practice using this table so that you become adept. Don't forget to apply temperature correction factors and adjustment factors if there are over three current-carrying conductors.

Article 312, "Cabinets, Cutout Boxes, and Meter Socket Enclosures," has a lot to do with wire-bending space and construction specifications, so it is of interest primarily to manufacturers and listing agencies.

Box Fill

Article 314, "Outlet, Device, Pull and Junction Boxes; Conduit Bodies; Fittings; and Hand-Hole Enclosures," contains some information that is relevant to licensing exams and code compliance on the job, so it is recommended that you become very familiar with it. Two major topics are contained in Article 314, and you can expect licensing exams to take note of both of them. Noncompliance will result in overcrowded boxes. When you try to put the cover in place, after stuffing in the conductors and devices, either the cover won't go in place, or if it will, you will have to use so much force that the wire nuts will come loose, making for intermittent and arcing connections. Overfilled enclosures generate more heat and dissipate it less efficiently. Moreover, all that excess combustible material can ignite and produce enough additional heat to initiate a premises fire outside the enclosure.

You can avoid overfilling enclosures by complying with *NEC* box-fill rules. To do so, you need to understand how to perform the calculations.

Table 314.16(A), "Metal Boxes," contains two categories of information. Where no volume allowances are required, the maximum number of conductors size 18 AWG through 6 AWG is given for variously sized boxes. The more common application of this table, however, is to find the minimum volume of these same boxes. This is for when the volume is not marked on the box. The procedure is to match up the box volume with the information in Table 314.16(B), "Volume Allowance Required per Conductor," plus clamp fill, support-fittings fill, device or equipment fill, and equipment-grounding conductor fill. Each of these categories has specific rules as to how it is to be applied. For example, where one or more internal clamps, whether factory- or field-supplied, are present in the box, a single volume allowance in accordance with Table 314.16(B) is to be made based on the largest conductor in the box. No allowance is required for a cable connector with its clamping mechanism outside the box.

Table 314.16(B) provides cubic-inch allowance values for conductors in sizes 18 AWG (1.50 in.^{3}) to 6 AWG (5.0 in.^{3}).

For complex calculations with several conductors and devices, you can make up a worksheet showing the fill allowances for each item. Then total them up and see what size box is required. You are sure to have at least one exam question on box-fill calculations. If you remember how the two tables work together, the only problems boil down to locating the relevant sections and reading through the fill rules so that you will be able to come up with an accurate total.

Where conductors 4 AWG or larger that are required to be insulated are in a raceway or cable, the minimum dimensions of pull or junction boxes are to comply with the following:

- *Straight pulls.* The length of the box or conduit body is to be not less than eight times the trade size of the largest raceway.
- *Angle or U pulls or splices.* The distance between each raceway entry inside the box or conduit body and the opposite wall is to be not less than six times the trade size of the largest raceway in a row. This distance is to be increased for additional entries by the amount of the sum of the diameters of all other raceway entries in the same row on the same wall of the box. Each row is to be calculated individually, and the single row that provides the maximum distance is to be used.

These are the two sets of box-fill requirements. If you put some time into assimilating these rules to the point where you can apply them on an open-book basis, you'll become an expert in this area, and you can design that part of the installation for your coworkers.

TIP ***There are two different methods, outlined earlier, for figuring box fill. Which method is to be used? It depends on the size of the conductors.***

The remainder of Chapter 3 consists of Articles 320, "Armored Cable: Type AC," to 399, "Outdoor Overhead Conductors over 600 V." Each of these articles covers a specific type of cable or raceway. For some reason, the first 11 articles, which cover types of cable, are in alphabetical order, whereas the others, which cover types of raceways, are not in alphabetical order. If you keep this in mind, you will be able to find the information you need quickly. The *National Electrical Code Handbook* has a listing of articles at the beginning of each chapter, but the *National Fire Protection Agency (NFPA) Code Book* does not, so you have to consult the main Table of Contents at the front of the book. (Rapid Code navigation is facilitated if you know when to consult the Index and when to consult the Table of Contents.) Assuming that you have figured out how to quickly find the relevant wiring method article, the next step is to learn how to navigate within the individual articles.

Cable and Raceway Specifications

Each article is structured in the same way. For example, Part I, "General," begins with a section on scope, followed by definitions. Part II, "Installation," contains two very useful sections, "Uses Permitted" and "Uses Not Permitted." Other important sections are "Bending Radius" and "Securing and Supporting." (*Securing* means making the cable or raceway so that it can't wiggle, and *supporting* means making it so that it won't fall.) All these articles have parallel internal numbering systems, making this part of the Code quite user-friendly.

We'll look at some of these articles, not in the order in which they appear in the Code, but instead beginning with the most common and progressing into relatively unknown territory.

Everyone is familiar with Romex cable. This is the trade name for Type NM cable, which designates nonmetallic-sheathed cable. Variants are Type NMC and Type NMS. It is the wire of choice for residential work because it is less expensive than alternatives, easy to install, and requires no raceway. It is covered in *NEC* Article 334. Type NM cable consists of insulated conductors enclosed within an overall nonmetallic jacket. Type NMC is essentially the same, but the outer jacket is corrosion-resistant. Type NMS also resembles Type NM, but the jacket also encloses signaling, data, and communications conductors.

All these cable types can be used for either exposed or concealed wiring. Theoretically, you can run Type NM exposed on the inside finish wall, but for dwellings, this is not done for aesthetic reasons. Even in a garage or shed, Romex is generally run as concealed wiring for appearance as well as affording better protection. If it is run exposed, it must not be subject to physical damage (Figure 5-6).

Part II lists "Uses Permitted":

- One- and two-family dwellings and their attached or detached garages and storage buildings.
- Multifamily dwellings permitted to be of Types III, IV, and V construction except as prohibited in Section 334.12, "Uses Not Permitted."

What's all this stuff about construction types?

At one time, the Code did not permit the use of Type NM cable and its variants in buildings three stories or higher. This has changed, and now the delineation has to do with construction types. They are covered in the Code's "Informative Annex E." The Informative Annexes are found after Chapter 9, outside the Code's main body. As the name implies, they are not mandatory rules but are provided solely for informational purposes. As such, they aid in interpreting the Code mandates. You should read through and understand these annexes. It is not necessary or possible to commit all this information to memory, but it is available for reference as needed.

Construction Types I through V have to do with the extent to which buildings are fire-resistant. Type I is the most fire resistant, and Type V is the least fire resistant.

FIGURE 5-6 • Metal plate required to protect cable closer than 1¼ in. from the edge of a framing member.

- Other structures permitted to be of Types III, IV, and V construction, except as prohibited in Section 334.12. Cables are to be concealed within walls, floors, or ceilings that provide a thermal barrier of material that has at least a 15-minute rating, as identified in listings of fire-rated assemblies.
- Cable trays in structures permitted to be Type III, IV, or V, where cables are identified for the use.
- Types I and II construction, where installed within raceways.

Still Struggling

The construction types, based on the fire resistance of a building, determine whether Type NM (Romex) cable can be used. Most electricians use Type EMT and Type MC for all commercial and industrial work. Type NM is almost always used in residential applications.

Type NM is permitted for both exposed and concealed work in normally dry locations, except where prohibited in Section 334.10(3). It is permitted to be installed or fished in air voids in masonry-block or tile walls.

Type NMC is permitted for both exposed and concealed work in dry, moist, damp, or corrosive locations, except as prohibited by Section 334.10(3).

Further clarification is contained in Section 334.12, "Uses Not Permitted." This list is not complete because there are other sections of the Code that further restrict the use of Types NM, NMC, and NMS cable. These types are not permitted:

- In any dwelling or structure not specifically permitted in Sections 334.10(1), (2), and (3)
- Exposed in dropped or suspended ceilings in other than one- and two-family and multifamily dwellings
- As service-entrance cable
- In commercial garages having hazardous locations
- In theaters and similar locations
- In motion-picture studios
- In storage-battery rooms
- In hoistways or on elevators or escalators
- Embedded in poured cement, concrete, or aggregate
- In hazardous locations

Note that in commercial garages having hazardous areas, use of Types NM, NMC, and NMS cable is prohibited anywhere in the garage, even if it is outside the area that is classified as hazardous.

These cables are permitted, as we have seen, in noncommercial attached or detached garages that go with dwellings. Their use in garages that are used by backyard mechanics is a frequently seen Code violation. If vehicles that are fueled by flammable liquid or gas are to be serviced and repaired, then the building is not to be wired in Romex.

"Uses Permitted" and "Uses Not Permitted" are frequently referenced when it comes to Romex. If there are gray areas or locations where interpretations may vary, it is always better to use a metal raceway, such as EMT. Many electricians wiring anything but one- and two-story, one- and two-family dwellings leave the Romex back in the shop.

Article 330, "Metal-Clad Cable: Type MC," covers this interesting and robust alternative to Type NM. It is easy to cut and terminate, and its use allows you to get beyond the limitations of Type NM. Moreover, it is easy to transition back and forth between Type MC and Type EMT, making the two in combination an ideal medium for all but the most restrictive environments, in which only rigid metal conduit (RMC) and its close relative, intermediate metal conduit (IMC), will do.

MC stands for *metal clad*. In outward appearance, it resembles Type BX, which was widely used at one time. There are three kinds of Type MC cable based on the outer metal sheath: interlocked metal tape, corrugated metal tube, and smooth metal tube. In some cases there is a nonmetallic jacket over the metal sheath.

TIP ***To transition between EMT and MC, use a 4- × 4-in. metallic box.***

The *NEC* defines Type MC cable as a factory assembly of one or more insulated circuit conductors with or without optical-fiber members enclosed in an armor of interlocking tape or a smooth or corrugated metal sheath.

Part II, "Installation," as always, includes sections titled "Uses Permitted" and "Uses Not Permitted." Uses Permitted are

- For services, feeders, and branch circuits
- For power, lighting, control, and signal circuits
- Indoors or outdoors
- Exposed or concealed

- To be direct buried where identified for such use
- In any raceway
- As aerial cable on a messenger
- In hazardous locations and embedded in plaster finish on brick or other masonry, except in damp or wet locations
- In wet locations where any of the following conditions is met:
  (a) The metallic covering is impervious to moisture.
  (b) A moisture-impervious jacket is provided under the metal covering.
  (c) The insulated conductors under the metallic covering are listed for use in wet locations, and a corrosion-resistant jacket is provided over the metallic sheath.
- Where single-conductor cables are used, all phase conductors and, where used, the grounded conductor are to be grouped together to minimize induced voltage on the sheath.

Uses Not Permitted include

- Where subject to physical damage
- Where exposed to any of the destructive conditions in (a) or (b), unless the metallic sheath or armor is resistant to the conditions or is protected by material resistant to the conditions
  (a) Direct buried in the earth or embedded in concrete unless identified for direct burial.
  (b) Exposed to cinder fills, strong chlorides, caustic alkalis or vapors of chlorine or of hydrochloric acids.

Minimum-bending-radius figures are given for Type MC cable. They vary depending on the type of sheath. You will notice that a fairly large radius has to be maintained so that the cable does not kink.

Still Struggling

Type MC cable is permitted for services, but it is rarely used in that application because PVC conduit and Type SE cable are less expensive and easier to use.

I have noted, in the discussion of Article 250, "Grounding and Bonding," that certain metallic raceways and metal sheaths qualify as equipment-grounding conductors. This is generally not true of Type MC. This is for the most part a moot point because the commonly used Type MC contains a green insulated wire that is intended to be hooked up as an equipment-grounding conductor. Some applications, however, such as within patient-care areas in a health-care facility, require redundant grounding. For this use, the nearest thing to Type MC that will qualify is Type AC, which looks about the same. The difference is that Type AC contains a metal strip that is in continuous contact with the inside of the metal sheath. In conjunction with this strip, the metal sheath qualifies as a grounding means. Aside from areas where redundant grounding is required, though, Type MC is used extensively, even in some of the less demanding hazardous areas.

Type MC sheathing may be cut using a hacksaw, but this procedure is difficult, and you can end up cutting into the insulated conductors inside. It is better to use a Roto-Split or MC cutter (also called a *BX cutter*). You put the armored cable into the tool's channel, squeeze the handle to hold the cable in place, and turn the crank, which is attached to a small circular saw blade with metal-cutting teeth. Give the unwanted piece of armor a twist and slide it off, leaving 8-in. lengths of conductors.

Using the correct connector together with a locknut, terminate the cable in an appropriately sized enclosure knockout, and you're done. An insulating anti-short bushing, called a *red head*, is put in place to protect the wires from the sharp cutoff sheathing (Figure 5-7). There is a very small opening in the connector so that an inspector can be sure that it is being used. Snap-in connectors are quick and easy to use and don't require the red head or locknut.

MC cable is easy to use and requires just the one tool, as described earlier.

As I noted, Type MC cable is frequently used along with Type EMT cable. Their "Uses Permitted" and "Uses Not Permitted" are very similar, and they are

FIGURE 5-7 • Red heads required for standard Type MC terminations.

seen together almost everywhere in commercial and industrial settings. To transition, mount a 4- × 4-in. box (or larger subject to your box-fill calculations), run the MC and EMT to it, terminate connectors in knockouts, and splice the conductors using wire nuts or, for larger sizes, split bolt connectors. EMT is a little more rugged, and if run straight and plumb with well-made bends, it has a more professional appearance. For drilled holes in studs, hollow cavities, or where close, difficult angle bends are needed, MC is the better choice.

EMT in smaller sizes can be easily cut with a hacksaw. In larger sizes or where numerous cuts are to be made, a portable handheld band saw is perfect. Article 358 covers EMT (Figure 5-8). It is an unthreaded thinwall raceway of circular cross section designed for the physical protection and routing of conductors and for use as an equipment-grounding conductor when installed using appropriate fittings. EMT is generally made of steel (ferrous) with protective coatings or aluminum (nonferrous).

PROBLEM 5-4

How can EMT be bent without danger of kinking it?

SOLUTION

A hand or power bender must be used. Any other method will kink the tubing, substantially reducing the inside diameter at the bend.

FIGURE 5-8 • EMT couplings come in various sizes. If they have set screws like those shown here, they are for dry locations only.

EMT, properly speaking, is not a conduit. It is tubing, and the correct term is *metal raceway*. However, when it is being installed, it is common to hear electricians speak of "putting the wire in conduit." And although it is referred to as *thinwall*, the raceway affords substantial protection against physical damage (Figure 5-9). For interior work, the couplings and connectors are held together with set screws, so there are cases where the raceway is observed to pull apart. Running a green equipment-grounding conductor helps to prevent loss of grounding continuity.

Generally, the EMT will not pull apart unless it has been inadequately secured and/or the building undergoes structural failure. Where greater protection is required, RMC can be used, but it is much more expensive and far more difficult to install. The thickwall conduit resembles galvanized water pipe, and the threaded joints will not come apart.

Part II, "Installation," contains sections on "Uses Permitted" and "Uses Not Permitted."

"Uses Permitted" indicates that

- The use of EMT is permitted for both exposed and concealed work.
- Ferrous or nonferrous EMT, elbows, couplings, and fittings are permitted to be installed in concrete, in direct contact with the earth, or in areas

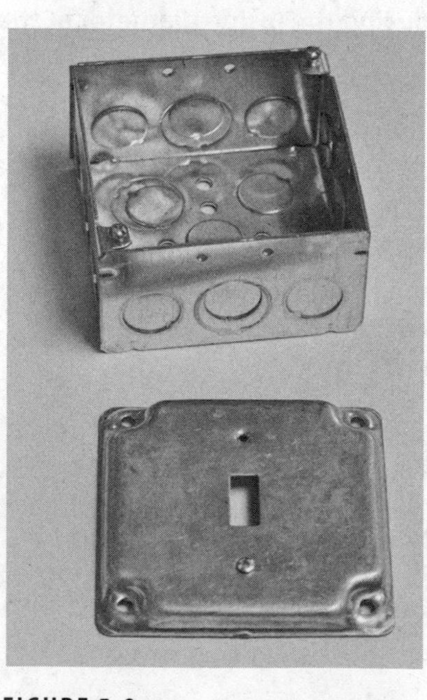

FIGURE 5-9 • A 4- × 4-in. metal box will accept EMT with the proper connector.

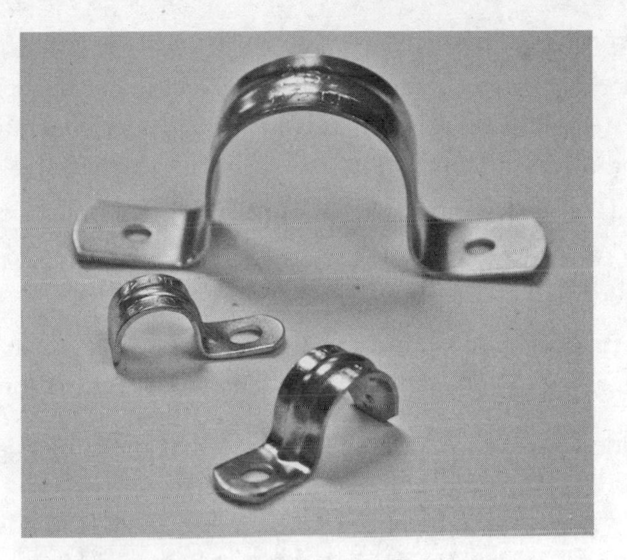

FIGURE 5-10 · Securing hardware for EMT.

subject to severe corrosive influences where protected by corrosion protection and approved as suitable for the condition.

- In wet locations, all supports, bolts, straps, screws, and so forth are to be of corrosion-resistant material or protected by corrosion-resistant material (Figure 5-10).

Despite the fact that EMT can be used, with conditions, embedded in concrete and underground, these applications are almost universally ceded over to PVC, which, in addition to being less expensive, is ideally suited for these uses. Where mounted outside on a wall, including for services, both EMT and PVC are used commonly. For long horizontal runs, it's better to stay away from PVC because it tends to buckle and deform with changes in temperature. In short vertical segments, as for services, PVC is used widely. It must always be gray Underwriters Laboratory (UL)–listed PVC conduit intended for electrical work, never white PVC water pipe or black ABS water pipe. This includes all fittings.

When EMT is deployed in a wet area, compression couplings, fittings, and connectors are needed. Setscrew hardware is for indoors only.

"Uses Not Permitted" lists these conditions:

- Where, during installation or afterwards, it will be subject to severe physical damage

- Where protected from corrosion solely by enamel
- In cinder concrete or cinder fill where subject to permanent moisture unless protected on all sides by a layer of noncinder concrete at least 2 in. thick or unless the tubing is at least 18 in. under the fill
- In any hazardous location except as permitted by other Code articles
- For the support of luminaires or other equipment, except conduit bodies no larger than the largest trade size of the tubing
- Where practicable, dissimilar metals in contact anywhere in the system are to be avoided to eliminate the possibility of galvanic action

It is provided that bends are to be made so that the tubing is not damaged and the internal diameter of the tubing is not effectively reduced. It is further provided that there are to be not more than four quarter bends (360 degrees total) between pull points, for example, conduit bodies and boxes (Figure 5-11). If this number is exceeded, it will be difficult, if not impossible, to install the conductors without damaging them. The answer is simple: install extra pull points as needed.

Additional requirements address securing and supporting intervals and can be referenced as needed.

We have reviewed three of the wiring methods that are employed in dwellings and nondwellings. There are many others in this series of articles. Some, such as integrated gas spacer cable, Type IGS, you may never have occasion

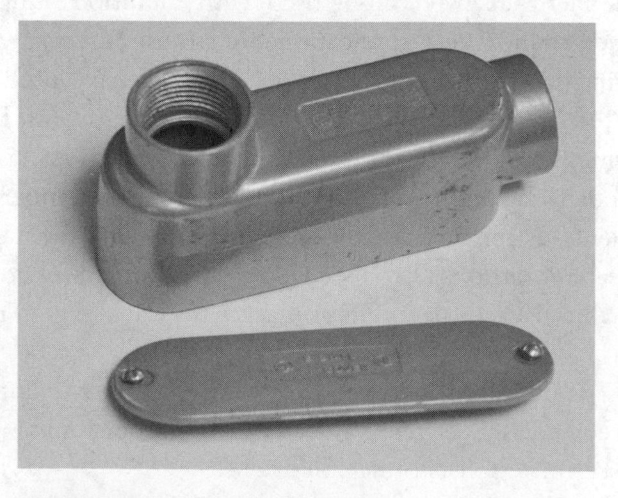

FIGURE 5-11 • Conduit LB used to facilitate pulling conductors when there are more than four 90-degree bends in a single segment of a run.

to use. Concealed knob-and-tube wiring may be encountered in a historical restoration project. Cable trays are a necessary part of some industrial and communications installations. Many premises building projects will include service-entrance cable, Types SE and USE. Licensing exams and on-the-job installations will require reference to these articles. The good news is that they are easy to find and well organized internally, conforming to a consistent template. Your experience with this part of the Code will be free of any difficulty if you put in the time.

This completes our survey of *NEC* Chapter 3, the final part of the Code that applies generally. Next on the agenda is a series of chapters that apply to specific types of equipment, locations, and conditions pertaining to all sorts of electrical work. It is beyond the scope or intent of this book to cover every detail. It is hoped, however, that after completing our survey, the reader will be equipped to find Code requirements quickly and easily.

QUIZ

These questions are intended to test your comprehension of Chapter 5. The passing score is 70 percent, but try to answer them all correctly. The quiz, like most electricians' tests, is open-book, so feel free to refer to the text. Answers appear in Answers to Quizzes and *NEC* Practice Exam.

1. **The purpose of the *NEC* is**
   A. the practical safeguarding of persons and property from hazards arising from the use of electricity.
   B. to train inexperienced individuals in performing electrical installations.
   C. to aid in creating an efficient end product.
   D. to provide design specifications for electrical installations.

2. **The *NEC* covers installations in both public and private buildings.**
   A. True
   B. False

3. ***NEC* Article 110 covers**
   A. conductor sizes.
   B. wiring integrity.
   C. interrupting rating.
   D. All of the above

4. **Grounded conductor and grounding conductor**
   A. are two names for the same thing.
   B. may be white or green.
   C. are at the same potential.
   D. are generally considered obsolete.

5. **All grounded conductors are identified by a continuous white finish from one end to the other.**
   A. True
   B. False

6. **In a multiwire branch circuit, the two ungrounded conductors**
   A. are connected to the same phase.
   B. are connected to opposite phases.
   C. originate from different panels.
   D. may be different voltages with respect to ground.

7. **GFCIs**
   A. are used in dwellings only.
   B. are never used outdoors.
   C. work by comparing the current in the hot wire to the current in the neutral.
   D. require an equipment-grounding conductor to work reliably.

8. A continuous load is expected to operate

A. for over an hour.
B. for over 2 hours.
C. for over 3 hours.
D. for as long as it takes to bring the heating element up to its maximum temperature.

9. A feeder always has an overcurrent device at each end.

A. True
B. False

10. The general lighting load for a dwelling is

A. 1 VA/ft^2.
B. 2 VA/ft^2.
C. 3 VA/ft^2.
D. dependent on the type of lighting.

chapter 6

National Electrical Code *Chapters 4 through 7: Getting Down to Specifics*

The remaining *National Electrical Code* (*NEC*) chapters, unlike those that went before, are quite specific. For this reason, it is easier to navigate within them and find the information you need. As I have stressed, in electrical work as in other types of building construction, time is of the essence. Whether you are in a timed exam, in conference with a customer, or at work on an installation, you need to get on with the task at hand, and that will not happen if you are thumbing through the Code in the hope of coming across the relevant material.

CHAPTER OBJECTIVES

In this chapter, you will

- Examine *NEC*'s coverage of equipment for general use.
- Learn about special occupancies, including hazardous locations.
- Find out what the *NEC* considers "special equipment."
- Learn about remote control, signaling, and power-limited equipment.

In this chapter, I certainly will not be able to cover the vast subject of electrical installation in all its complexity. The intent is to provide the reader with a sense of the overall structure of the body of knowledge in order to lay the foundation for competency and finally proficiency in this extensive field.

Equipment for General Use

Chapter 4, "Equipment for General Use," contains 21 articles, each covering a specific type of electrical equipment. This chapter is similar in content to Chapter 6, "Special Equipment," so one of your tasks as an electrician is to get a sense of which equipment is covered in each of these two complementary chapters. You can do this by comparing the lists of articles they contain, found in the Table of Contents. Notice that Chapter 4 equipment, as its title implies, is more general—switches, appliances, motors, and the like—whereas Chapter 6 equipment is more likely to be the final product that the end user sees—electric signs, elevators, and solar photovoltaic (PV) systems. Think of it this way: Chapter 4 covers the motor, whereas Chapter 6 covers the elevator that contains the motor.

The first article, 400, "Flexible Cords and Cables," provides information on these unique conductors. They have characteristics, described in this article, that make them stand apart from other wiring media. If you don't look carefully at the requirements, you may make a mistake that will be costly or even hazardous. In residential and commercial or industrial work, fire and shock are common results of the misapplication or faulty installation of flexible cords. In an ideal world, flexible cords would be kept short and few, but the reality is that in many homes and businesses they are overused to the point where a dangerous accumulation behind desks and under furniture generates excessive heat. Poor connections are in close proximity to combustible material, and bare copper strands are exposed for little fingers to find, setting the stage for shock for those who are most vulnerable.

These hazards are mitigated by closely adhering to the Code mandates. Article 400 starts off with a multipage table that lists and describes the many types of recognized flexible cords and cables, giving the range of sizes, number of conductors, insulation types, and permitted uses. Another table, 400.5(A)(1), "Allowable Ampacity for Flexible Cords and Cables," lists these many types, with differing ampacities depending on whether there are two or three current-carrying conductors. Key parts of this article are "Uses Permitted" and "Uses Not Permitted."

Uses Permitted include

- Pendants
- Wiring of luminaires
- Connection of portable luminaires, portable and mobile signs, or appliances
- Elevator cables
- Wiring of cranes and hoists
- Connection of utilization equipment to facilitate frequent interchange
- Prevention of transmission of noise or vibration
- Appliances where the fastening means and mechanical connections are specifically designed to permit ready removal for maintenance and repair and the appliance is intended or identified for flexible-cord connection
- Connection of moving parts
- Where specifically permitted elsewhere in the Code

Uses Not Permitted include

- As a substitute for the fixed wiring of a structure
- Where run through holes in walls, structural ceilings, suspended ceilings, dropped ceilings, or floors
- Where run through doorways, windows, or similar openings
- Where attached to building surfaces
- Where concealed by walls, floors, or ceilings or located above suspended or dropped ceilings
- Where installed in raceways
- Where subject to physical abuse

Every one of these "Uses Not Permitted" is a frequently seen Code violation, and numerous building fires can be traced to these misuses.

Article 402, "Fixture Wires," is similar in content to Article 400. Fixture wires are used primarily to connect luminaires to the branch-circuit conductors. As such, they are generally a part of the factory assembly, not field-supplied. They are usually stranded, although some types may be solid.

TIP ***When repairing old light fixtures, it is often helpful to use solder to tin the ends of stranded fixture wire so that they will go under the terminal screws.***

Article 404, "Switches," covers the installation and use of these ubiquitous devices. Beginning electricians and many home owners know exactly how to wire a common single-pole snap switch in part because it is so simple—just a two-wire item in series with the load—and in part because it is such a common task. A small residence will have over a dozen switches, not counting the circuit breakers.

Despite their simplicity, there are some subtle points that may cause difficulty or be completely missed. If your jobs are subject to electrical inspection, you can be sure that the authority having jurisdiction (AHJ) will check on a requirement that appeared in *NEC 2011*. It provided that where switches that control lighting loads are supplied by a grounded general-purpose branch circuit, the grounded-circuit conductor for the controlled lighting circuit is to be provided at the switch location. For a switch that feeds power to the load, the requirement has no effect because there will be a grounded conductor at the switch location anyway. But in the case of a switch loop, it was customary to bring power first to the load location where the grounded conductor would be connected to the correct terminal of the load. Then the ungrounded conductor, instead of being similarly connected to the load as in the case of an unswitched setup, would be connected using wire nuts to an ungrounded conductor going to the switch. A separate ungrounded conductor would return and be connected to the ungrounded terminal of the load. These two wires are known as a *switch loop*. If run in raceway, they could both be black, or one could be black and one red or any color other than white or green. If run in Romex, a white wire would have to be used for one of the ungrounded conductors, and it would have to be reidentified by marking it black at both ends.

Under the new rule, a grounded conductor has to be run to the switch. At the enclosure, the end must be long enough to comply with the free-conductor provisions in Chapter 3. It is to be coiled up and made available for future use. What is the purpose of this strange requirement? The thinking is that in the future the owner or tenant may wish to install an electronic lighting-control device as an energy-saving strategy. A proximity sensor can be installed in the wall box, and the light fixture will go on when a person enters the room. A device of this sort requires a small amount of power to operate, so it needs the grounded conductor.

Most apprentice electricians understand that the single-pole switch, like other control devices, goes in the ungrounded (black) conductor in series with the load. If wired in the grounded (white) conductor in series with the load, the switch would be effective in breaking the circuit and shutting down the load. But it would be a Code violation because even though the load would be powered down, one terminal would remain hot, and this could be deceptive and a shock hazard for a person working on the piece of equipment or wiring. The same applies to a circuit breaker or a fuse.

Three- and Four-Way Switches

There are several additional Code requirements that can be found in Article 404, but there isn't too much more to be said about the devices except in regard to three- and four-way switches. Home owners successfully wire a garage or addition and do a credible job of it but become confused when it comes to wiring a pair of three-way switches, and they have to call in an electrician to sort things out. Often a miswired pair of three-way switches stems from an improperly connected neutral.

You won't go wrong if you remember a few simple principles. The purpose of a three-way switch installation is to permit control of a load, usually a ceiling light fixture, from two separate locations. If it is possible to enter a darkened room from two separate entrances at opposite ends, three-way switches are essential. If there were a pair of standard single-pole switches, one at each end of the room, there would be a problem. If the switches were in series, they would both have to be on to turn on the light. If they were in parallel, they would both have to be off to turn off the light. Either way, there would be instances where the occupant would have to cross the darkened room to control the light. Three-way switches permit the occupant to control the light from either location. A standard single-pole switch has a set ON position, normally at the top. A three-way switch, in contrast, changes the status of the load when it is thrown, regardless of position. This is so because between the 2 three-way switches there are two ungrounded conductors that are alternate hot legs. Electricians call them *travelers*, and by operating either one of the three-way switches, you can control the load as desired.

The pair of three-way switches may be regarded as a black box or single device that acts like a standard single-pole switch. The difference is that the standard single-pole equivalent thought-experiment switch could be located in only one place, whereas the real-world pair of three-way switches can be located in two separate locations. What makes this possible is the three-conductor cable that connects the 2 three-way switches.

TIP ***If you forget how to wire a three-way switch, ring it out with your ohmmeter to find which terminal is the input or output and which terminals are the travelers.***

To wire a power source, a pair of three-way switches, and load successfully the first time, keep these points in mind: There are several configurations. In the simplest version, the power source is connected to the first of the three-way switches. The power flow is then to the second three-way switch and finally to the load. As mentioned earlier, the 2 three-way switches function as one standard single-pole switch.

A three-way switch (Figure 6-1) gets its name from the fact that it has three terminals, one at one end of the switch body and two at the other end. The single terminal is labeled "common." Two conductors, one grounded and one ungrounded, are brought from the entrance panel or other power source to the first three-way switch enclosure. The live ungrounded (black) conductor is connected to the common terminal at the first three-way switch. The grounded (white) conductor is not connected to the switch. Instead, it is connected by means of a wire nut to the white ungrounded conductor that goes to the second three-way switch enclosure. In other words, it just passes through the enclosure without being connected to the device.

A specialized type of cable runs from the first to the second three-way switch. It has three conductors, one grounded and two ungrounded. If you are

FIGURE 6-1 • A three-way switch. The two terminals at one end of the switch body are for the travelers.

using Romex, it will be Type NM 14-3. [Lots of electricians use 12 American Wire Gauge (AWG) wire for most premises branch circuits, even where 14 AWG would suffice to comply with *NEC* Chapter 3 ampacity tables. The reason in part is to simplify inventory. But it is customary to use Type NM 14-3 for the run between three-way switches. The lighting load is minimal, often less than 1 A.] The Type NM 14-3 will have one white, one black, and one red conductor. If you are doing a raceway installation, it will be essentially the same, but the colors for the ungrounded conductors differ. The two ungrounded conductors connect to the two terminals that are at one end of each three-way switch. Again, the white is connected by means of a wire nut to a white conductor that exits the second three-way switch box, runs to the load, and connects to the grounded terminal. Note that the white conductor runs from the neutral bar of the entrance panel or other power source and goes directly to the load without being connected to any switching device.

As for the cable from the second three-way switch to the load, in a dwelling, it is usually Type NM 12-2. (Throughout this discussion, I make no mention of the equipment-grounding conductor. It is assumed that it will be terminated correctly at the power source, both switch enclosures, and the load. I also do not repeat the extra neutral that must be brought to the device when it is a switch-loop situation.)

Because you know that the 2 three-way switches are equivalent to a standard single-pole switch, the hookups at the second three-way switch and load will be simple.

If you keep the preceding principles in mind, wiring a power source, a pair of three-way switches, and load will not present a problem. Just remember that the whites run straight through, never connecting to a switch. The 12-2 ungrounded conductor connects to the first three-way switch common terminal, constituting the input of the black box, and another 12-2 ungrounded conductor connects to the second three-way switch common terminal. The black and red conductors that are part of the 14-3 cable run only between the 2 three-way switches, connecting to the two terminals at each switch that are not marked "common."

That's all there is to it, as far as the straight power-feed configuration is concerned. Because of the geography of the room and where the power source is relative to the switches and luminaire, it may be more economical to run the power from the source directly to the load. If this is the case, the wiring details are a little different. If you are using Romex, Type NM 12-2 goes from the power source to the light fixture. The grounded (white) conductor connects to the appropriate terminal. That takes care of the neutral. There are no other neutrals. Because this is a switch-loop installation, all the remaining whites are ungrounded (hot) conductors and as such have to be reidentified by marking them black at all terminations. The black conductor that comes from the power source goes inside the light fixture enclosure, then connects to a black conductor that is part of a 12-2 cable, and connects to the common terminal of what I'll continue to call the second three-way switch. The white conductor, reidentified as black at both ends, connects to the ungrounded terminal at the load and inside the second three-way switch enclosure connects to the white conductor of the Type NM 14-3. That white conductor is now an ungrounded part of a switch loop, and as such, it is reidentified black. It runs to the first three-way switch and connects to the common terminal, which now has become the black-box output.

Still Struggling

Always connect the travelers in the three-conductor cable (or raceway) to the two terminals of each three-way switch that are not labeled "common."

Another configuration is seen when power from the source runs, via Type NM 12-2, to the second three-way switch. I'll leave you to figure out that one. Here's a hint: One white wire is a neutral ungrounded conductor, and the other, part of the 14-3 cable, is part of a switch loop and as such is reidentified black.

Still more elaborate configurations are possible when there are more than one light fixture with three-way switches located (electrically) between them. They all work according to the same principles. It's primarily a matter of keeping in mind whether a white is a neutral or part of a switch loop. If it is a neutral, it is connected only at the power source and load, perhaps passing through a switch enclosure but never connected to the switch. It may be a part of the 12-2 or a part of the 14-3.

If the white is part of a switch loop, it is reidentified black at both ends. One end is connected to live power from the source, and the other end is connected to a common terminal of a three-way switch.

If the load is to be controlled at more than two locations, the way this is done is simpler than you may think. Any number of four-way switches may be inserted between the 2 three-way switches. A four-way switch has four terminals, two at each end of the switch body. A red and black pair is connected at each end of the four-way switch. It doesn't matter which terminals are input and which are output. Nor does it matter which terminal is red and which is black. If they become reversed any number of times in the course of connecting to the four-way switches between the 2 three-way switches, there will be no effect on the operation of the circuit. So we see that a four-way switch is easier to wire than a three-way switch.

PROBLEM 6-1

What are the maximum numbers of three-way and four-way switches in a single circuit?

SOLUTION

In a single circuit, there will never be more than 2 three-way switches. The number of four-way switches is not limited.

A Common Violation

Article 406, "Receptacles, Cord Connectors and Attachment Plugs (Caps)," contains provisions relating to the safe installation and use of these devices. An example of the kind of rules found here is that a receptacle is not to be installed within or directly over a bathtub or shower stall. Moreover, receptacles are not permitted to be installed in a face-up position in countertops or similar work surfaces.

A common Code violation and very hazardous practice is to connect an attachment plug via a cord to a generator or other power source so that it can be plugged into a receptacle to backfeed premises wiring during an outage. The prongs of an attachment plug are to be live only if plugged into a receptacle or cord connector so that they cannot be contacted.

There are guidelines regarding replacement of grounding-type receptacles and installation of ground-fault circuit interrupters (GFCIs), arc-fault circuit interrupters (AFCIs), and tamper-resistant receptacles. When these devices first became mandatory in prescribed locations, it was not anticipated that they would be immediately installed in existing locations. Instead, they are to be used to replace faulty standard receptacles as needed.

Article 408, "Switchboards and Panelboards," covers these types of equipment. A *switchboard* is defined in Chapter 1 as a large single panel, frame, or assembly of panels on which are mounted on the face, back, or both. It contains switches, overcurrent or other protective devices, buses, and usually instruments. Switchboards are generally accessible from the rear as well as the front and are not intended to be installed in cabinets.

A *panelboard* is defined as a single panel or group of panel units designed for assembly in the form of a single panel, including buses and automatic overcurrent devices, and equipped with or without switches for the control of light, heat, or power circuits; designed to be placed in a cabinet or cutout box placed in or against a wall, partition, or other support; and accessible only from the front. These two items are similar in purpose but different in outward appearance, as described earlier.

Electricians spend a lot of time working on the service-entrance panel during initial installation and in the course of subsequent maintenance, troubleshooting, and repair. Because the entrance panel is fed from a utility transformer or other "stiff" power source, and the conductors are large with high ampacity, there is a great potential for arc-flash incidents. When working on an entrance panel, it should be disconnected from the power supply. This can be done by opening the main disconnect if it is in a separate upstream enclosure. Another method for powering down the service-entrance panel if it is supplied by a utility is to remove the meter from the meter socket. This leaves live supply-side terminals in the meter socket. The utility will probably give you a plastic shield that goes inside the meter socket cover, taking the place of the meter. The problem remains, however, that you have removed the utility's seal, so the cover and shield can be removed by a curious child. Furthermore, most of the time the meter is not within sight of the electrician working on the entrance panel. Depending on the area, it might be prudent to post a guard at the meter socket and to discuss the matter with a utility representative. Of course, because it is the utility's equipment, the meter cannot be pulled without getting the permission of the utility. This is a routine matter and is usually taken care of with a simple phone call. Under no circumstances outside of an emergency should the meter seal be removed without utility consent.

Occupational Safety and Health Administration (OSHA) regulations state that live equipment is not to be worked on except under certain limited conditions, which are spelled out. Among other things, these conditions include use of protective clothing and shielding, the extent of which varies with the amount of hazardous available fault current.

Electricians are frequently asked to change the entrance panel in a residence, that is, take out the old one and put in a new one. This is done in the course of an upgrade, changing to a larger service, or to replace an old fuse box. Sometimes, as a result of an addition or renovations, electrical usage increases, and it is decided to go, let us say, from a 100-A to a 200-A service. Of course, this will mean a new box, with the appropriately sized main breaker. Also, larger service conductors and in some cases a larger ground electrode conductor will be needed (Figure 6-2). Also, the 200-A entrance panel is physically larger, so available space has to be verified.

Before pulling the meter, get everything ready. Because power will be out during the changeover, you will need cordless tools. Also, lighting usually becomes an issue, especially if the entrance panel is in a basement with no natural light. You'll need to set up battery-powered lighting, or if possible, you can run a cord from another building. Doing an entrance panel change-out by flashlight is not enjoyable.

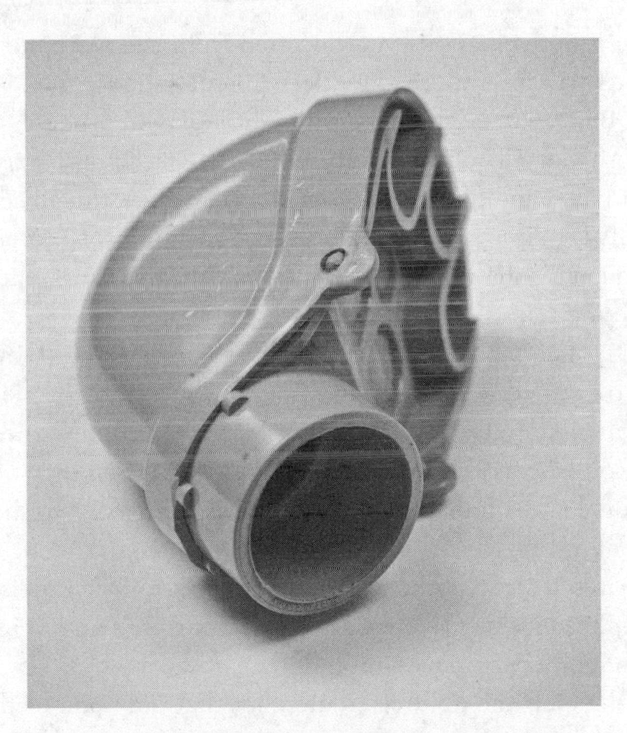

FIGURE 6-2 • An upgrade to a 200-A service will involve new service conductors and weatherhead. Always put the neutral through the middle hole. Mount the weatherhead high enough so that the service drop will have sufficient ground clearance.

FIGURE 6-3 • A 4- × 4-in. metal box with offset nipple that is helpful in positioning adjacent enclosures. Sometimes this setup is needed if the cable will not quite reach the new panel location.

If the new panel is in a different location from the old panel, many of the branch circuits will have to be extended. A large junction box will be needed (Figure 6-3). Be sure to calculate box fill. This box will probably be larger than you expect. An elegant solution is to remove breakers, busbars, and hardware from the old panel and use it as a splice box. This has the advantage that the branch circuits are already in place. However, such improvised usage may void the listing. In such situations, speak in advance to the AHJ, who has the final word on such matters.

If the new panel is going in the same location as the old one, the branch circuits can be pulled out of the old box and put into the new one. When removing branch-circuit wires from the old panel, unscrew the locknuts and remove the Romex connectors along with the wires. Then they can be installed in the knockouts of the replacement box. This is as opposed to sliding the wires out of the connectors, which is more difficult and risks scraping bits of insulation off the wires. If some of the wires are too short to reach their terminations, they can be spliced inside the entrance panel using wire nuts (Figure 6-4), or if necessary, junction boxes can be mounted nearby.

When hooking up the neutrals, remember that only one of the white conductors can be put in each terminal opening. This is so because at some time in the future a worker may need to back out the screw to remove a neutral, and in the process, the required ground continuity to another circuit would be interrupted. In contrast, it is permissible to put more than one bare or green

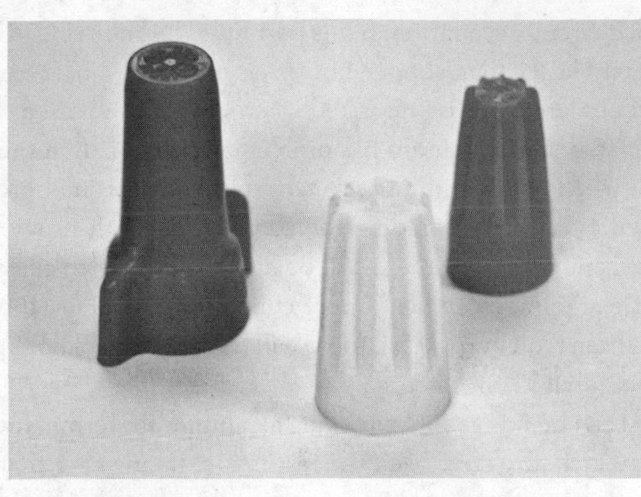

FIGURE 6-4 • Wire nuts, technically known as *solderless connectors*, are excellent for splicing wires inside enclosures. If alterations to the wiring are needed, the wire nuts can be easily twisted off and reused.

equipment-grounding conductor under a single screw in the equipment-grounding terminal bar (Figure 6-5).

Before changing over the branch circuits, it is best to wire the service-entrance conductors into the main breaker. The reason for doing this first is so that these large wires will be routed along the back of the panel, not blocking access to the branch-circuit wiring. The same is true of the grounding-electrode conductor and the heavy wire that is used to bond any metallic water pipe that may be in the building. (Don't neglect that one!)

If you are going to a larger service, remember that you have to put in larger service-entrance conductors. If it is a 120/240-V, three-wire, single-phase service for a dwelling only, reduced conductors sizes are permitted, as shown in Table 310.15(B)(7). Generally, electricians do not use aluminum wire in small

FIGURE 6-5 • Knock-out blanks are used to plug unwanted openings in an enclosure.

sizes for branch circuits even though it is permitted by the *NEC*. Aluminum has a lower ampacity rating than copper, but this is not really a problem because for each application a larger size is prescribed where aluminum is used. The difficulty lies elsewhere. It is that aluminum has problematic termination properties. Specifically, even when an aluminum wire termination is torqued properly, it will seem to loosen over time, degrading the connection. With heavy current flow relative to the conductor size, there will be heat, giving rise to corrosion, a more resistive electrical joint, more heat, more corrosion, and so on, the result being heavy arcing. If the fault burns clear, there will be an outage. Before that happens, there is the potential for an electrical fire.

This hazard can be mitigated by making all aluminum terminations correctly. The procedure is first to strip back the insulation, taking care that no particles remain. Then thoroughly clean the mating metal surfaces with a stiff steel wire brush. Next, apply a coating of corrosion inhibitor to the wire and lug surfaces. Finally, using a torque screwdriver or wrench, tighten the lug to the specifications given by the manufacturer. Obviously, it is not feasible to go through all these steps for each of the hundreds or thousands of electrical splices and terminations in a dwelling, not to mention in commercial or industrial work.

Notwithstanding, aluminum wire is used in electrical services because in the large sizes involved, the cost of copper wire is excessive. Moreover, there are relatively few terminations, so here aluminum has the advantage. If you have to do a short segment such as straight through the wall stub from meter socket to entrance panel or where conduit fill may be a problem, consider copper.

A final point: In going to a larger-size aerial service, the conductors from the supply side of the meter socket up to the weatherhead will have to be larger as well, again conforming to Table 310.15(B)(7). Upstream from the point of connection, the wiring is done by utility workers, and it is not under *NEC* jurisdiction. Conductor sizes may be smaller in part because they are in free air and there is better heat dissipation.

Before putting the meter back into the socket, check your work with an ohmmeter. Also, shut off the main breaker. There have been instances of utility workers energizing a service by putting in the meter with no one in the house. Fires have been traced to an electric cook stove inadvertently left on with paper and construction materials in contact with the burners. In this connection, it is good practice always to fully torque all terminations when they are first made.

Article 410, "Luminaires, Lampholders, and Lamps," and Article 422, "Appliances," contain provisions intended to promote safety for these types of equipment.

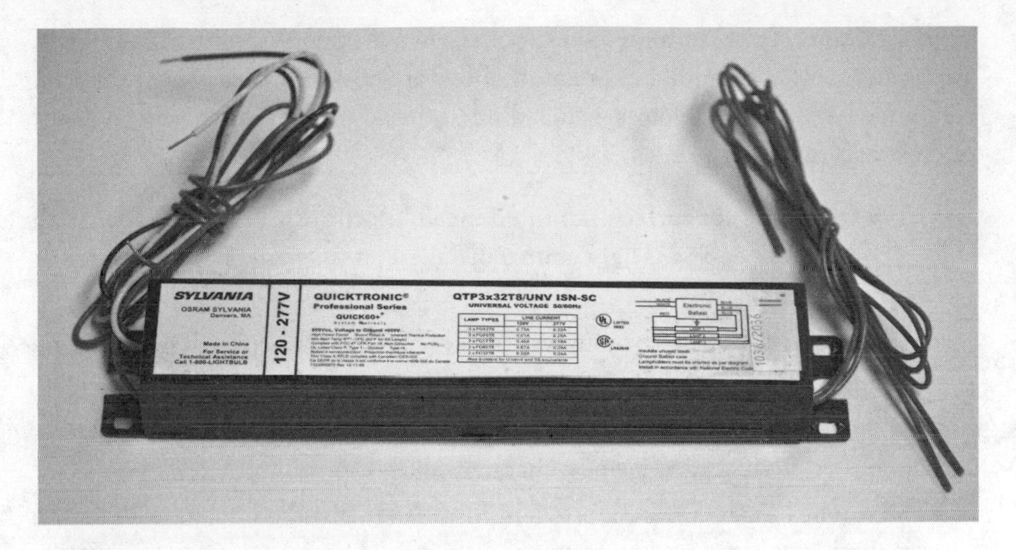

FIGURE 6-6 • Replacement ballast for fluorescent light fixture. Electricians install them frequently.

They generate heat in proportion to the power they consume, and this heat has to be dissipated in a manner that will not cause a temperature rise to the point where nearby combustible material could ignite (Figure 6-6).

Design and construction of luminaires are critical. Testing organizations such as Underwriters Laboratories (UL) examine electrical equipment from the point of view of safety. Property owners, electricians, and electrical inspectors rely on product listing as far as the internal wiring is concerned. As for installation, *NEC* guidelines contain provisions that ensure that temperature rise, loose wires, and other defects will not become issues. These two articles are not likely to be the focus of a great many exam questions, but because luminaires and appliances are so much a part of an electrician's work, it is essential to hear what the Code has to say about them.

Luminaires in clothes closets can be a definite fire hazard because there is usually a concentration of combustible material. Also, the area is confined with limited air circulation, so the temperature can rise. There are two problems that have emerged in regard to the luminaire. It radiates heat in normal operation, and if there is an equipment malfunction, the temperature may rise still higher. An incandescent bulb, if it breaks, can release a hot filament that may ignite combustible material below. Fluorescent luminaires run cooler and are generally less hazardous, although if the current through the ballast increases above the normal level, the ballast can overheat. Light-emitting diode (LED) lamps run cool and are reliable, but there is still the possibility of wiring error.

In a clothes closet, incandescent luminaires with open or partially enclosed lamps and pendant luminaires or lampholders are not permitted. The minimum clearance between luminaires installed in clothes closets and the nearest point of closet storage space is

- Twelve inches for surface-mounted incandescent or LED luminaires with a completely enclosed light source installed on the wall above the door or on the ceiling.
- Six inches for surface-mounted fluorescent luminaires installed on the wall above the door or on the ceiling.
- Six inches for recessed incandescent or LED luminaires with a completely enclosed light source installed in the wall or the ceiling.
- Six inches for recessed fluorescent luminaires installed in the wall or the ceiling.
- Surface-mounted fluorescent or LED luminaires are permitted to be installed within the closet storage space where identified for this use.

Appliance is defined in Article 100 as utilization equipment, generally other than industrial, that is normally built in standardized sizes or types and is installed or connected as a unit to perform one or more functions such as clothes washing, air conditioning, food mixing, deep frying, and so forth.

Article 422 contains specific provisions for various types of appliances, and electricians must be familiar with them in order to do the installations correctly. As an example, a storage-type hot water heater that has a capacity of 120 gal or less is considered a continuous load, and as such, its overcurrent device and conductors are to be sized based on 125 percent of the appliance nameplate rating.

There are requirements for central heating equipment, infrared-lamp industrial appliances, central-outlet vacuum assemblies, and ceiling-suspended (paddle) fans, among others. Three articles on specific types of heating equipment follow, and if you are installing baseboard electric heat in a residence or a boiler employing resistance-type immersion heating elements, you will want to review the material to ensure that the installation is done correctly.

Motors

This takes us to Article 430, "Motors, Motor Circuits, and Controllers." This *NEC* article is over 50 pages long (in the 2011 handbook edition), and there is a large amount of information to assimilate. If you have been doing power and light wiring in dwellings and suddenly find yourself working in a heavy-industrial setting, Article 430 will be your guide. In this sort of location, motors are everywhere. Depending on the type of production, some of these motors are likely to be large three-phase machines, and there may be variable-frequency drives (VFDs). All of these have to be installed, maintained, diagnosed, and repaired.

One way that the Code comes in is sizing the conductors, disconnects, and overcurrent protection. This is done in quite a different way for motors than for other types of electrical equipment for the simple reason that motors draw much more current when starting than when up to speed and running. All electrical equipment, even a light bulb, is like this, but motors are more so.

I'll explain the *NEC* motor sizing procedures in a while, but first we should look at the various types of motors and see what makes them run. An *electric motor*, broadly defined, is a device that converts electrical energy to mechanical motion. This motion can be lineal, as in the solenoid in a Delco Remy automotive starter that causes the rotating gear to engage with the flywheel. Even such a commonplace item as a loudspeaker is actually a motor, although it is more commonly thought of as an actuator or transducer. Most motors that electricians install and maintain are rotary machines, and these are the devices that I will be discussing.

Almost all motors depend for their operation on the interaction between two magnetic fields or on the interaction between one magnetic field and a magnetically permeable material such as soft iron. When electric current flows through a conductor, a magnetic field is established around the conductor, as discovered quite by accident when the Danish physicist Hans Christian Oersted observed in 1829 that the needle of a compass was deflected when brought near a live wire. This magnetic field is weak, barely able to move the compass needle. But it was soon discovered that if the conductor were wound in a coil with many turns, particularly around a magnetically permeable material, the magnetic field would be greatly increased. Taking advantage of this phenomenon, early experimenters were able to build motors that would actually rotate, although at first they were mere scientific curiosities not able to perform useful work.

Early experimenters had access to electrical power only from inefficient chemical batteries. Real working motors did not become available until Thomas Edison, Nikola Tesla, and George Westinghouse built their competing electrical generation and distribution systems, which enabled the use of electric motors in the factory and on the farm. Edison came first, and his direct-current (dc) power gave rise to powerful dc motors.

All motors work the same, but there are some significant differences in the circuitry. A motor consists of an outer housing. Mounted on the inside are coils that are connected to the power supply or, alternatively, to permanent magnets. At either end of this housing, centered in the two-end bell housings, are bearings that hold the rotor securely in place yet allow it to turn freely. Attached to the rotor is a shaft that extends outside the housing. (Sometimes there are two shafts, one at each end. Sometimes there are no external shafts, as in a rotary inverter.) The shaft can have an attached pulley, gear, saw blade, grinding wheel, or other tool so that useful work can be accomplished.

The rotor also has coils mounted on it or some variation. It is the magnetic interaction between rotor and stator that makes for motor rotation.

Perpetual Motion

At times, various schemes have surfaced whereby permanent magnets would be mounted in both rotor and stator, creating a perpetual-motion machine that could do useful work without requiring outside power. Such a machine could never function. It would violate the law of conservation of energy. At best, after assembly, the rotor would rotate a part turn and thereafter remain at rest. The motor would work on a continuous basis only if the polarity of one of the permanent magnets were to be reversed as the motor turned. The only way I can think of to reverse this polarity is by continuously physically flipping the magnet end to end. This would require more energy than the output of the motor.

? Still Struggling

You can have permanent magnets in the rotor or in the stator but not in both. Power from the supply has to be fed into either the rotor or the stator, and it has to be continuously switched as the motor turns.

Returning to the dc motor, in its simplest form, electrical energy from the outside dc power source is fed to both stator and rotor. The stator can be wired directly to the power supply. The rotor cannot be so wired. If it were, the wires would quickly twist and break off. The dilemma is how to get power into the rotor. The solution is to feed the electric current through brushes that ride over a segmented commutator that is part of the rotor and turns with it. At one time, these were actually copper brushes, but solid-carbon bars were found to perform better. The name persists. The brushes must exert a slight, not excessive, pressure on the commutator. The ends where contact is made are ground to a concave shape to match the curvature of the commutator so that there is good electrical contact with a minimum of sparking and wear.

The brush-commutator assembly serves another purpose as well. As the rotor turns, the brushes contact different commutator segments, whose output is polarized differently. This switching makes possible the turning of the rotor on a continuous basis. The magnetic field established by the stator remains stationary, whereas the magnetic field established by the rotor is constantly being switched in time with the turning of the rotor. The magnetic field associated with the rotor is constantly chasing after the magnetic field associated with the stator, and constrained by bearings, the rotor is pulled around in a circular fashion.

That is how a dc motor works. The scheme is known as *internal commutation*, and because the switching is inherent in the brush-commutator assembly, the speed of the motor can be controlled solely by the voltage of the power supply. Direction of rotation can be controlled by simply reversing the polarity of the dc motor's input. Even after the introduction of alternating-current (ac) motors, dc motors continued to predominate for elevators and similar applications where speed and direction of rotation must be varied.

Ac motors were developed by Tesla, and their use became widespread as Westinghouse gained market share. Like dc motors, ac motors require switching of the magnetic fields with respect to one another. In the ac motor, the commutation is external, and motion occurs because the electrical supply changes polarity in a regular, ongoing manner. There are two major types of ac motors, synchronous and asynchronous. In a *synchronous* motor, the speed of the motor is exactly synchronized with some integral multiple (depending on the number of poles in the rotor) of the ac line frequency. In an *asynchronous* (induction) motor, the speed also depends on the ac line frequency, but it is not exactly synchronized. It is always slower by a certain percentage, and this speed reduction is known as *slip*.

Returning for now to the synchronous ac motor, there are several possible configurations. The rotation of the shaft is synchronized with the frequency of the electrical supply. The rotation period equals an integral number of ac cycles. There are electromagnets built into the stator, creating a magnetic field. The magnetic field of the stator rotates in accord with the line-current pulses, and the rotor turns along with this field at the same speed.

Synchronous motors are available in a wide range of sizes depending on the application. Very small synchronous motors are used for clocks and timers. Because the speed is locked into the line frequency, they are accurate to a very high standard. Utilities keep track of their output, and if an aberration in frequency is detected, they temporarily alter the speed of the generator a very slight amount until the error is corrected. Small synchronous motors are self-excited, meaning that they will start on their own.

TIP ***Before wiring a motor, read the nameplate carefully because it contains information needed to correctly size the circuit.***

Very large industrial-scale synchronous motors are not self-excited because they have too much inertia of rest. Some other solution must be chosen to start them, such as an auxiliary induction pony motor or induction windings embedded within the rotor. The large synchronous motor is more expensive than its induction rival, but it has distinct advantages. Besides being accurately synchronized to the line frequency, it is more efficient than the induction motor. Moreover, it operates at leading or unity power factor, providing power-factor correction for the facility in which it is located.

Tesla's extensive research into alternating current and polyphase power led him to invent the induction motor. Electric current and magnetism are developed in the rotor by means of induction. In effect, the stator is the primary of a standard ac transformer, and the rotor is the secondary. There is no direct electrical connection to the rotor and therefore no brushes nor commutator. It is an elegant solution, inexpensive, low maintenance, and long lasting. Because of these advantages, perhaps 90 percent of the motors in an industrial facility are induction motors.

What Is Slip?

As in a synchronous motor, the ac supplied to the induction motor's stator creates a magnetic field whose rotation conforms to the line frequency. This induces electric current in each of the rotor's windings. As a result, there are magnetic fields in the rotor that cause the rotor to turn as they chase after the ever-receding magnetic fields associated with the stator. If the rotor turned at synchronous speed, there would no induced current in the rotor windings and no magnetic field. The rotor must always turn a certain amount slower than the stator's rotating field. The difference in speed is known as *slip*. It must be emphasized that the rotor is not merely out of phase with the stator's magnetic field but actually turning at a slower rate. For this reason, induction motors are not used to power clocks or timers.

Still Struggling

Slip is to be understood not as some kind of lost or wasted motion but as a necessary part of the process that allows the induction motor to work.

I stated that the speed of a dc motor can be precisely and smoothly regulated by varying the supply voltage. For an ac motor, synchronous or induction, this is not true. For these motors, the speed is frequency-dependent, not voltage-dependent. To a certain extent, you can slow down an ac motor by reducing the voltage, but this is equivalent to loading it more heavily, and it is not a good way to control speed because more heat is produced, resulting in shortened motor life.

In the last few decades, electronic control systems have been developed that permit us to vary the speed of an ac motor. These devices are known as *variable-frequency drives* (VFDs), *adjustable-frequency drives*, *variable-speed drives*, *adjustable-speed drives*, *ac drives*, *inverter drives*, and so on. All these terms are equivalent and may be used interchangeably, except that some are applicable to nonelectric systems, as in the field of hydraulics. This nomenclature is sometimes used to include the motor, more often just the drive.

Most VFDs are used with three-phase induction motors. Synchronous motors are preferable in some situations, and single-phase motors are possible, but the usual setup is for a VFD to control the speed of a three-phase induction motor. Any existing ac motor can be fitted up with a VFD with this precaution: If the motor has an internal fan provided for cooling purposes, when motor speed is reduced, the cooling will be reduced as well, setting the stage for overheating. Of course, it is always possible to augment the motor's cooling system or find a way to reduce the ambient temperature.

The VFD assembly consists of the motor, the main drive controller, and the operator interface. The controller, in turn, has these subcomponents: the bridge rectifier, a dc link, and the inverter. The operator interface permits a human to start, stop, and adjust the speed of the motor. Numerous additional monitoring and control functions are possible. Depending on the application, it may be necessary for the operator to reverse the direction of rotation or switch between various modes.

The operator interface typically will include an alphanumeric display, indicator lights, gauges, and a keypad. A serial communication port permits connection to a laptop so that the VFD can be configured, modified, and monitored using proprietary software.

When first starting the motor, the VFD supplies low voltage and frequency, which are ramped up until the desired speed is achieved. This process avoids the high inrush current experienced during conventional line operation. Many VFDs incorporate bypass circuitry so that the motor can be fed directly off the line when the operator chooses to run the motor at its rated speed.

When the system fails to operate as intended, the electrician must commence troubleshooting. As always, the first step is to check VFD input and output. For a three-phase motor, there is the possibility of loss of one phase. When this happens, the motor will run poorly, overheat, and sustain damage.

A large motor will draw a significant amount of current, increasing as the motor ages, accompanied by a rise in temperature. A small temperature rise can be tolerated, but it should be investigated before it becomes problematic. Heat may come from a bearing or the windings. A thermal imaging instrument is valuable in monitoring the state of the motor. The time to start doing this is when the system is operating normally. Keep records so that damaging trends can be detected before there is an outage and expensive downtime.

When there are motor problems, always suspect the load. If it binds up or runs out of true, this can look like a motor problem. Belt drives, conveyers, mixers, pumps, and other types of loads can run poorly and lead you to believe that the motor or even the VFD is at fault. Ordinary troubleshooting techniques will lead you to the bad system or component.

If there are normal voltages at the VFD input but not the motor input, disconnect the power supply and open the VFD cabinet. Beware of hazardous voltages that may persist long after the unit is powered down. Electrolytic capacitors store powerful electric charges. Some technicians short these devices with jumper wires, but it is better to discharge the stored energy by shunting power resistors of appropriate rating across the terminals.

Look for defects in the transformer, diodes, or capacitors. These devices can be checked with an ohmmeter. First, check that there is no power on the dc bus. Switch the ohmmeter to the diode check function, and touch the negative probe to the positive dc bus. Touch the positive probe to each of the input conductors that go to the rectifier diodes. This reverse biases them so that you read a voltage drop at each input terminal. Next, put the positive lead on the negative dc bus. Place the negative lead on each input line to ascertain whether there is a forward diode drop. By doing these readings, you can find out if the diode bridge is open, shorted, or good.

If the input rectifier diodes are okay, check the VFD output section. Connect the positive lead to the negative dc bus, and place the negative lead on each of the three output terminals that are connected to the motor. There should be a low forward-biased voltage drop. You may see that the output circuitry is shorted or open, as opposed to acting like a semiconductor. A possibility also is that the dc bus fuse is blown. This could point to a different fault within the enclosure.

Check all input and output devices. If any of them appear burned or distorted, the VFD will not work. Check the capacitors, both visually and with the multimeter. An oscilloscope will determine if there is too much ripple at the rectifier output.

These diagnostic procedures will verify or evaluate the main current path through the VFD. There is an entire additional area that makes problems for many technicians. For the VFD to work properly, operating parameters have to be programmed into the machine. The VFD behaves as expected when it receives commands from the user interface.

The user interface may have diagnostic capability, and if so, it will be able to display error codes on the alphanumeric display. The error codes should be listed in the user manual, with instructions for remedial steps required to clear them. VFD makes and models vary, but all can be fixed, or they aren't broken! If problems persist and the solution appears elusive, it is suggested that you install a telephone jack near the VFD so that you can consult the manufacturer's tech helpline while you are at work on the equipment.

TIP ***The VFD can be diagnosed using simple troubleshooting techniques, usually with a multimeter.***

Stepper Motors

Another type of motor is the *stepper*, variously called a *stepmotor* or a *stepping motor*. It differs from other types of motors in that it is able to spin in either direction or turn part rotations in an incremental fashion. It can stop with or without holding torque, and these properties make it valuable in a great many seemingly unrelated applications.

The amount of each increment is the *step angle*. The maximum number of steps in a complete rotation comprises the *resolution*, and this varies depending on the model. The step angle could range from 1 to 90 degrees. The stepper motor turns a single incremental step and then stops, whether or not voltage is maintained. If current continues to flow through the winding, the stepper does not turn, but there is a pronounced holding torque.

The stepper is simple, inexpensive, and very reliable. There are no brushes or commutator. All commutation takes place outside, and the signal fed to the motor is generated electronically. An audio frequency could be applied to the input terminals, and the stepper motor would turn at a speed determined by the frequency of the signal and the resolution of the motor. Or it could be made to turn one step per hour or one step per day or at any rate determined by the controller. The controller will itself have inputs so that it can make the motor perform in accordance with internal programming based on the state of one or more sensors or an input from a human operator.

There are many applications—inkjet printers, disc drives, dot-matrix printers, fax machines, or wherever precise and rapid positioning is required. Relatively large stepper motors were found in the old large-dish satellite TV units that had to point to different satellites in the sky as channels were changed. Those satellite dishes are obsolete now, so the stepper motors can be had for the asking. Another application is for moving tracking-type solar photovoltaic (PV) arrays so that they can be kept facing the sun as it moves across the sky.

Stepper motors are ubiquitous in the emerging field of robotics, and we can expect to see more of them as this equipment proliferates.

For a stepper motor, failures are rare. Because there are no brushes or commutator, the only parts that can malfunction are the bearings and the coils. Bearing failure is infrequent because of low revolutions per minute (rpm), a forgiving duty cycle when operated in stepping mode, and light loading. The coils are not generally subjected to overvoltage because of overcurrent protection in the controller and protective diodes to prevent inductive voltage spikes due to fast switching.

There are several types of stepper motors. They are similar in purpose and operating characteristics, differing only in some mechanical and electrical details in ways that are easy to understand.

Still Struggling

Stepper motors and small dc motors may not have nameplates. They are identified by counting the leads.

The variable-reluctance stepper motor has a stator and one moving part, the rotor. As in other rotary motors, the rotor is mounted on front and rear bearings. The rotor consists of a soft-iron notched-tooth disk. The notched teeth are the essence of the whole thing, and they are what makes this kind of motor work. Because soft iron is more magnetically permeable than air, the teeth, where they approach closer to the stator coils, are attracted to them when the coils are energized by pulses from the controller. This makes the rotor spin or turn incrementally in response to those pulses. The stator consists of field windings. Generally, there are three, four, or five of them. If there are no markings or nameplate on the motor, the number of leads will help to identify it. There will be one wire for each coil plus one common wire, so if there are four coils, there will be five wires. It is easy to ring them out with an ohmmeter in order to find the common wire. If it is a small stepper motor, you can manually pulse it with a 9-V battery.

The common wire is usually connected to the positive pole of the dc power supply, and the negative voltage is applied by the controller to each winding in sequence. A series of pulses, sometimes very rapid, makes the stepper motor perform as required. A variable-reluctance stepper motor can be identified by the fact that when no power is being applied, it will turn very freely by hand. The permanent-magnet stepper, when turned by hand, will exhibit a pronounced cogging.

Unlike the variable-reluctance stepper motor, the permanent-magnet stepper motor works because of attraction and repulsion between the permanent magnet mounted on the rotor and windings mounted on the stator. Both these parts of the motor have sets of teeth, and they are offset with respect to one another. When timed pulses from the controller are sent to the stator windings, the rotor will turn incrementally.

There are two types of permanent-magnet stepper motors, unipolar and bipolar. The *bipolar* stepper motor has one winding per phase on the stator. The *unipolar* stepper motor is the same, except for the fact that each winding has a center tap. Unipolar stepper motors can function as bipolar models if you disregard the center taps and connect nothing to them.

The bipolar stepper motor is less complex, but its controller requires more circuitry. To reverse a unipolar stepper motor, you switch each section of windings.

Because bipolar stepper motors have a single winding per phase, to reverse rotation, it is necessary to reverse the magnetic poles. This is accomplished within the controller by means of an H-bridge.

Hybrid stepper motors are more costly. They incorporate features of both the variable-reluctance and permanent-magnet steppers. The hybrid stepper motor has better speed, torque, and step resolution. There can be as many as 400 steps per rotation! The rotor has many teeth, and there is also a permanent magnet.

If you are interested in carrying out experiments or building a prototype, or if you need a replacement, stepper motors (as well as small two-wire dc motors) may be taken out of discarded ink-jet printers. Because a new computer often comes with a printer, there are plenty of surplus inkjet printers available. Be sure to save the drive transistors, mounting hardware, LEDs, and so on that are associated with them.

Working with Three-Phase Circuits

Three-phase power is a mysterious, unknown area to many home owners and beginning electricians because it is not found in any but the largest residences. But most more experienced electricians, especially maintenance workers in commercial and industrial settings, install and repair three-phase wiring and distribution and end-user equipment on a daily basis (Figure 6-7). Once you learn the fundamental concepts of three-phase power, you will find it user-friendly and entirely workable given the large amounts of current that are often conveyed.

Where three-phase power becomes most relevant is in the supply of motors. It is not practical to build single-phase motors over 5 hp, and they are very rare. At 1 hp and over, the advantages in wiring a three-phase motor are palpable. In many industrial facilities, even fractional-horsepower three-phase motors are used, if only because they are less expensive initially.

PROBLEM 6-2

Why are all motors not three-phase?

SOLUTION

In many buildings, such as dwellings and small commercial establishments, three-phase power is not available because the service is single-phase. Moreover, in rural and small residential areas, three-phase power is not available on the street.

FIGURE 6-7 • A three-pole circuit breaker is used to originate a three-phase branch circuit at a three-phase panel. It picks up power from the three busbars.

A good way to begin learning about three-phase power is by driving through an industrial area that has aerial distribution lines and service drops and observing the wiring and connections. You can stand safely outside the grounded metal fence surrounding a substation and visually trace the wiring—incoming three-phase power from the grid, outgoing three-phase power for local distribution, single-phase power tapped off for use in the associated maintenance building or for yard lighting, and so on.

As you drive through neighborhoods with three-phase aerial distribution, notice how the wiring is configured. There will be three conductors at the highest level. The heights and distances between them depend on the voltage, and these specifications are laid out in the *National Electrical Safety Code*, which has jurisdiction over utility generation and distribution lines. In long-range distribution lines, rather than three conductors, there may be six, run in parallel to get better ampacity while maintaining manageable wire sizes.

A few feet below these hot lines is a single conductor, the grounded neutral. The reason it is run at a lower level is so that if it becomes detached from the insulators and falls, it will not lay across the phase lines, causing a line-to-line fault. Also, the grounded neutral is less dangerous than the phase lines if hit by a downed tree.

Below the neutral are communication and data lines, including telephone, cable television (CATV), and broadband lines. Utility workers who maintain them should be able to work from aerial buckets without contacting the higher-voltage lines.

It is interesting to observe that when more than one building is connected to a single transformer, it is possible for the more remote building to be served by a 240-V single-phase neutral conductor that is also the neutral for a much higher voltage that is part of the main distribution line going in the opposite direction.

As you pass buildings that have three- and single-phase services, look at the transformer connections. The three-phase transformers have three primary conductors tapped off the distribution lines and three secondaries going via service drops to the buildings, where three conductors will be routed to the meter and then into the building to the entrance panel. (These final runs will be in conduit, so you won't be able to distinguish the individual conductors unless you happen to be around when a meter is being pulled or while the service is under construction.)

It is against the Code (with exceptions) to have more than one service at a building. Sometimes service drops that are made up of paralleled conductors can be deceiving in this regard.

Three-phase electrical systems are characterized by three hot conductors that have a voltage difference between one another. Each has a voltage potential with respect to the neutral. The three conductors each carry currents of the same voltage and frequency that are out of phase with one another. If you visualize the time period of the cycle as a complete circle, the waveforms of the three circuits are equally spaced 120 degrees apart. They are wired together in either of two configurations, wye or delta. To wire a motor or other load, including an entrance panel or load center, you don't have to be concerned with whether it is a wye or delta configuration, except in regard to one color-coding detail (discussed below).

Three-phase motors may be synchronous or asynchronous (induction). They are both wired the same. All you have to do is bring the three conductors to the motor and connect them to the three motor leads. It makes no difference which branch-circuit conductor is connected to which motor lead except in regard to motor rotation, load balancing, and for a high-leg delta, color coding. We'll discuss these concerns in that order.

For any three-phase motor, synchronous or induction, motor rotation can be reversed by interchanging the connections of any two leads. You can reverse these connections at any accessible location—entrance panel, junction box, disconnect, or at the motor. The only thing you have to remember is that if it is a group installation, the other motors also will be affected if the wires are reversed upstream of where they are tapped off.

In some applications, the motor rotation will be set at the original installation. Other times it will be necessary to install controls, either automatic or manual, so that the operator can change motor direction as needed. This could be accomplished by means of a four-way switch, but be aware that the switch has to be rated for the application.

Motor rotation can be determined by trial and error, except that some pumps can be damaged if run in the wrong direction.

Some pumps and air blowers will produce liquid or air flow in the same direction regardless of motor rotation, but with better output for one direction of rotation than the other. This is so because of the cup shape of the impellers or fan blades. You can try both rotations and see which works better. This is particularly true of a three-phase submersible well pump.

An alternate method is to make use of a three-phase motor-rotation meter. This instrument can be bought for around $100 and comes with operating instructions.

Phase balancing is another matter that you will want to consider, especially for large motors. This is a question of matching line and load phases without getting the motor rotation wrong. If you measure three-phase line voltages with equal loads connected, you will observe slight variations, usually under 3 percent of the nominal value. This could be because of utility imbalance, inconsistencies in branch-circuit connections or conductors, or unbalanced loading. Similarly, there can be variations in the motor winding impedances. These variations cannot be measured with an ohmmeter because the readings will not correspond to winding impedances under dynamic running conditions. The variations, like the line-voltage variations, will usually be slight, but depending on the connections, the properties can either reinforce or cancel out one another. So the object is to match line and load optimally without affecting the direction of motor rotation. The way to do this is by "rolling" the connections. Change A to B, B to C, and

C to A. In this way, rotation is never affected. For each combination, power up the motor with its normal load, and measure the current through one of the conductors using your clamp-on ammeter. When the three readings are almost nearly equal, you have made the final connections. Of course, this procedure should be performed after correct motor direction has been established.

The *NEC* provides that the insulation of any ungrounded conductor is to be any color other than white, which is reserved for grounded conductors, and green, which is reserved for grounding conductors. Even though not Code mandated, it is common electricians' trade practice to color-code three-phase wiring as follows:

- 120/208Y: black, red, blue
- 277/480Y: brown, orange, yellow (delta) or brown, violet, yellow (wye)

The *NEC* requires white or gray for the neutral and bare or green for the equipment-grounding conductor.

For a delta configuration, if the midpoint of one phase is grounded, the opposite leg will have a higher voltage with respect to ground. This high leg is to be color-coded orange.

A three-phase entrance panel or load center designed for circuit breakers is like a single-phase box except that there are three busbars instead of two. The three main conductors enter the enclosure and are connected to a three-pole main breaker, which supplies power to the three busbars. In a single-phase panel, the double-pole main breaker feeds the two busbars. In the three-phase box it is possible to get three-phase power or single-phase power. To pick up three-phase power, three-pole breakers are used. They connect to all three busbars, and they have three terminals so that the three-conductor branch circuit will be able to power a three-phase motor or other three-phase load. Single-phase power is available from this same panel (Figure 6-8). A double-pole breaker is used to pick up any two legs depending on its position in the box (Figure 6-9), or a single-phase breaker can connect to one busbar, and the single hot wire can make up a circuit in conjunction with a white wire connected to the neutral bar. The neutral bar, equipment-grounding bar, and main bonding jumper are all hooked up in the same way as in single-phase work. The meter socket has three input and three output lugs, and the wiring is self-evident, the grounding electrode conductor being connected as in a single-phase meter socket.

The only thing that is at all difficult in wiring three-phase power is sizing the conductors and overcurrent protection, and if this is not done correctly, you'll have a situation where you have installed a large amount of equipment and

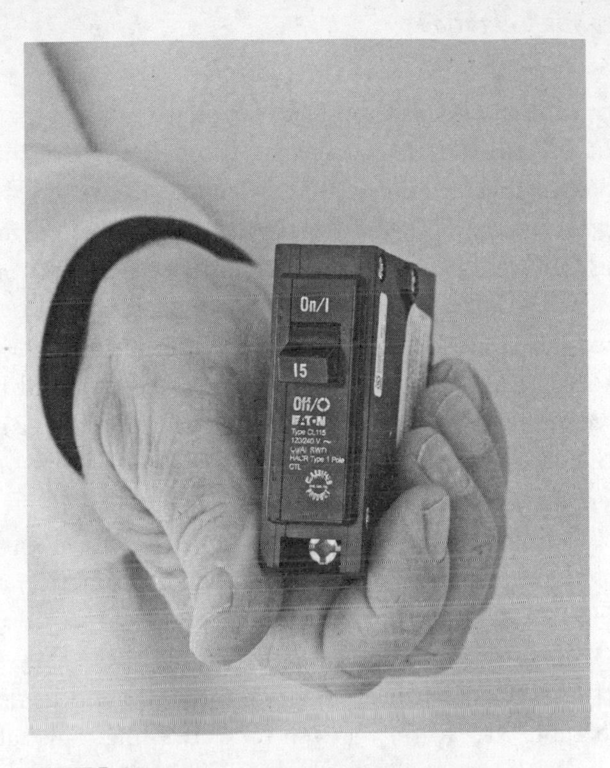

FIGURE 6-8 • Fifteen-ampere single-pole breaker used for 120-V branch circuits throughout a building.

FIGURE 6-9 • Twenty-ampere double-pole breaker used for light 240-V branch circuits, such as individual electric baseboard heating units.

wiring only to find that it keeps tripping the overcurrent devices, or worse, they do not trip, but the conductors overheat to the point where the insulation is damaged or combustible building materials ignite.

If it is a large commercial or industrial job, an electrical engineer may have sized everything in architectural drawings and specifications. Still, if you are installing the work, you will want to verify the calculations because anyone could make a mistake. In residential and smaller commercial work, it is customary for the on-site electricians to size the service, feeder, and branch-circuit conductors. If you are a journeyman electrician working under a master electrician, it is in your interest to become involved in the decision-making process and learn to perform the calculations.

If three-phase motor loads, especially group installations, are part of this picture, there is a new level of complexity, but with practice, all of this will become familiar. I will turn back to *NEC* Article 430 at this time to see how motor circuits differ from other electric circuits and how they are sized. The first thing to note is that in sizing three-phase service and feeder conductors, you cannot use Table 310.15(B)(7), even for a dwelling. This table is only for single-phase dwelling services and feeders.

To size ordinary nonmotor loads, the procedure is to use the full-load rating on the nameplate. Motors are different. If the ampacity of motor branch-circuit conductors, branch-circuit and ground-fault protection, and ampere rating of the motor disconnecting means were to be determined by the ampere rating on the motor nameplate, the motor would fail to start because the overcurrent device would cut out before the motor got up to speed (Figure 6-10). This is so because motors draw unusually high current in the first few seconds of operation, before they get up to rated speed. Just about all electrical loads are like this, but motors are more so.

In order to allow a motor to start and yet still provide adequate protection, a unique two-part overcurrent-protection setup has been developed. Short-circuit and ground-fault overcurrent protection is installed at the upstream end of the branch circuit, and overload protection is installed at the downstream end, adjacent to the motor. It is necessary, in wiring a motor circuit, to have both these devices in place, and it is also necessary that the devices have the correct ratings in amperes.

First, you will want to choose the value for the ampacity of conductors, ampere rating of switches, and branch-circuit short-circuit and ground-fault protection (Figure 6-11). Instead of using the current rating that is on the motor nameplate, the procedure is to take the horsepower rating of the motor

FIGURE 6-10 • Startup switch for motor-control circuit.

FIGURE 6-11 • Cartridge fuses provide reliable overcurrent protection.

off the nameplate and refer to the appropriate table to find the full-load current. These tables are found in the final part of Article 430. This is Part XIV, "Tables," and the information you now need to extract for any given motor is found in one of the following tables:

- Table 430.247, "Full-Load Current in Amperes, Direct-Current Motors"
- Table 430.248, "Full-Load Current in Amperes, Single-Phase Alternating-Current Motors"
- Table 430.249, "Full-Load Current in Amperes, Two-Phase Alternating-Current Motors (Four-Wire)"
- Table 430.250, "Full-Load Current, Three-Phase Alternating-Current Motors"

It is a simple matter to choose the applicable table and look up the full-load current. Most of the time you will be consulting either Table 430.248 or Table 430.250 for single-phase or three-phase ac motors. (Many large dc motors, for elevators and similar applications, are still in use, but for new installations they have been eclipsed for the most part by induction motors equipped with VFDs. Two-phase systems are nearly obsolete.)

These tables list, down the left-hand column, a range of horsepowers (for three-phase, ½ to 590 hp) and, across the top, voltages from 90 to 2,300 V. The three-phase table lists induction and synchronous motors separately.

Using these values, you can size the electrical supply to the motor. Then, refer to Part IV, "Motor Branch-Circuit Short-Circuit and Ground-Fault Protection." It is this section that permits much higher-rated overcurrent devices than you might expect. These devices, being higher-rated, offer protection only for ground faults and short circuits. They do not protect the motor from overload. That is done separately, as you shall see.

Part IV, Table 430.52, "Maximum Rating or Setting of Motor Branch-Circuit Short-Circuit and Ground-Fault Protective Devices," gives percentages of the full-load current that may be applied when selecting the overcurrent device. The percentages vary from 150 to 1,100 percent depending on the type of motor and the type of overcurrent device. Look this table over carefully because the information it contains is central to the whole process of sizing breakers or fuses that protect the branch-circuit conductors, switches, disconnects, and so on.

Once the short-circuit and ground-fault protection has been sized, you can move on to the other part of the equation, which is overload protection. Remember that this is a separate level of protection intended to protect the motor from overheating and being damaged. It also protects the branch-circuit conductors from temperature rise that might take place if the motor were to be made to work too hard as a result of excessive loading or if the motor failed to start.

Motor Overload Protection

Part III, "Motor and Branch-Circuit Overload Protection," covers this second level of protection. To begin, it is stated that overload protection is not required where a power loss would create a hazard, as in the case of fire pumps. A fire-pump motor could begin to sustain damage due to overload but should continue to run even up to the point of failure so that its firefighting function is not interrupted. Of course, for such equipment, the first level of protection, branch-circuit short-circuit and ground-fault protection, is still required. It is only the overload protection that may be omitted.

Methods for providing overload protection are varied and depend on the size and duty of the motor, whether it is automatically started, and other factors. A great many motors are used in continuous-duty applications and are greater than 1 hp. For these motors, the Code specifies four options:

- An overload device may be located in the motor controller, and this may be a so-called heater. At a certain temperature, a metal part of the linkage melts, causing the controller to interrupt current to the motor. After it cools, the device can be reset. Heaters are available in a range of ampere ratings compatible with different controllers. The device is to be selected to trip at certain specified percentages of the motor nameplate full-load current rating: motors with a service factor of 1.15 or greater, 125 percent; motors with a marked temperature rise of 40°C or less, 125 percent; all other motors, 115 percent. Service factor and temperature rise are found on the nameplate.
- A thermal-protector integral with the motor is another method of overload protection for this type of motor. The device can be separate from the motor, its control circuit operated by a protective device integral with the motor.
- A protective device integral with the motor that will protect the motor against failure to start is permitted if the motor is part of an approved assembly that does not normally subject the motor to overload.
- For motors larger than 1,500 hp, a protective device having embedded temperature detectors is another option for motor overload protection. They are to cause current to the motor to be interrupted when the motor attains a temperature rise greater than marked on the nameplate in an ambient temperature of 40°C.

These motor overload-protection methods are applicable to other motors as well. When wiring a motor, the procedure is to check the Code to determine which category the motor falls into. Then choose among the methods permitted. The motor nameplate full-load current is used to select the level of motor overload protection, not the full-load current from Tables 430.248 through 430.250, which are used for sizing the feeder and branch-circuit wiring.

For single-phase, 240-V motors, the overload device is required to be placed in only one ungrounded conductor. This is counterintuitive because switching and overcurrent protection is ordinarily installed in both hot lines. The purpose of the overload protection, however, is not to serve as a disconnect but rather merely as overload protection. The disconnect is provided elsewhere, upstream of the overload protection.

In the case of a three-phase motor, however, overload protection is required in all three legs because running with a dropped phase will cause a three-phase motor to overheat and be damaged.

Placement of overload devices is spelled out in Table 430.37, "Overload Units."

Part IX, "Disconnecting Means," requires that all motors are to have disconnecting means capable of disconnecting both motors and controllers from the circuit. The disconnect is to be located in sight of the controller.

This requirement is intended to protect an individual who may be working on the motor from shock or injury if the circuit should become energized unexpectedly. Serious injuries are possible, for example, when a belt-driven fan starts with the guard removed.

For stationary motors of 1/8 hp or less, the branch-circuit overcurrent device is permitted to serve as the disconnecting means.

For cord-and-plug-connected motors, a horsepower-rated attachment plug and receptacle may serve as the disconnecting means.

Refrigeration Equipment

Article 430 is followed by Article 440, "Air-Conditioning and Refrigerating Equipment." It must be emphasized that individuals working on this type of equipment are required by federal law in the United States to comply with Environmental Protection Agency (EPA) regulations that pertain to the handling of refrigerant. These rules are in addition to state and local licensing requirements, and penalties are very high.

Electrical refrigeration equipment is motorized, so Article 440 may be considered supplementary to Article 430, and both articles must be implemented together. The procedure for sizing the branch-circuit conductors and overcurrent protection for refrigeration is somewhat different from that for motors in general, and I will take this up later in this chapter, but first an overview.

The main focus in Article 440 is on hermetic refrigerant motor compressors. Special considerations apply to this equipment and for any air-conditioning or refrigeration equipment that is supplied from a branch circuit that supplies a hermetic refrigerant motor compressor. The hermetic refrigerant motor compressor is a very common configuration in refrigeration. At one time, refrigeration equipment consisted of a motor that powered a compressor by means of pulleys and a V-belt drive. The problem with this arrangement was that in time the refrigerant in the compressor would leak around the input shaft because of wear on the seal. This difficulty has been solved by building the motor and compressor as a single hermetically sealed unit in the same housing with no external shaft or shaft seals to worry about. The hermetically sealed motor-compressor is generally reliable, although in a demanding work environment, its life is not infinite. Damage can result from outside power surges, loss of phase for a three-phase unit, or water contamination of the refrigerant, creating an acidic mix that will etch through the insulation coating on the windings, grounding them out. The hermetic unit is not field-repairable.

For a walk-in cooler, the compressor is usually installed in a separate location, perhaps in a well-ventilated compressor room in the basement, grouped with similar units. The compressor receives its supply of refrigerant from a low-pressure line coming from the evaporator in the walk-in. Passing through the compressor, the refrigerant is changed from a gas to a liquid. Under pressure, it suddenly occupies a much smaller volume, and as a consequence, its temperature increases substantially. Now, if the pressure were decreased to its former level, its temperature would revert accordingly, and there would be no refrigeration. What makes cooling possible is what happens next. Immediately after leaving the compressor, the hot, compressed refrigerant passes through a heat exchanger that resembles an automotive radiator. (Water-cooled arrangements are also possible.) A motorized fan cools the heat exchanger, known as the *condenser*, from the outside, and now, still pressurized, the refrigerant is approximately room temperature. Via a small-diameter copper high-pressure line, it moves into the walk-in, where it passes through a second heat exchanger (Figure 6-12), again resembling an automotive radiator with a motor-driven fan. The airflow across this *evaporator*, as it is called, cools the box to the desired

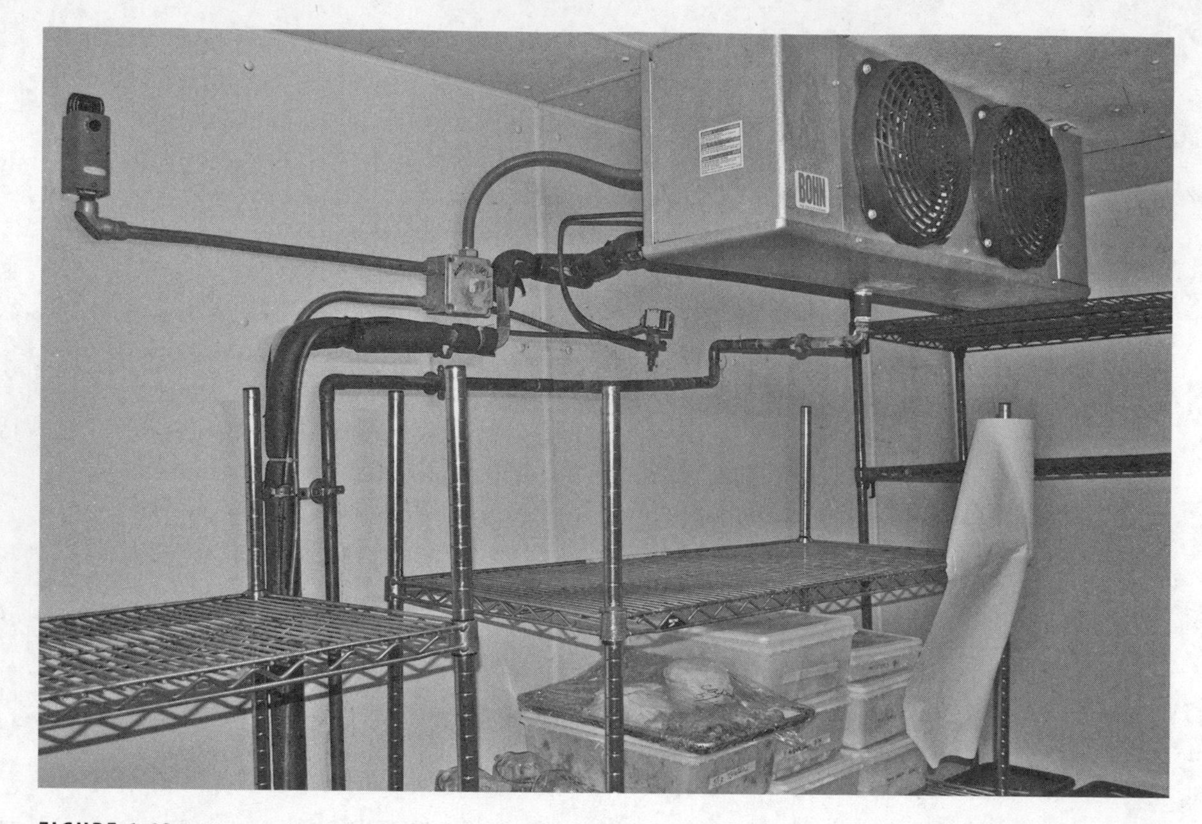

FIGURE 6-12 • Evaporator in a walk-in.

temperature, which is regulated by means of a simple thermostat. The system may be proportioned either for cooling or as a freezer.

When I was first exposed to these systems, I was perplexed as to how they could ever work. How could the room-temperature refrigerant passing through the evaporator result in cooling action? The answer is by means of a small, inconspicuous device known as the *diffuser valve*. It consists, in addition to a valve that is capable of regulating the system, of a small in-line orifice that purposely restricts the flow of refrigerant. It should be located as close as possible to the wall of the walk-in. It causes the refrigerant to revert to its gaseous state at a much lower pressure. Because per unit of mass the refrigerant now occupies a much larger volume, its temperature is considerably lower. It is very cold, and passing through the evaporator, it is capable of cooling the entire walk-in. (The walls of the walk-in are well insulated, as is the door, which has good weather stripping.)

This is basically how any refrigeration or air-conditioning system works, although there are numerous auxiliary controls, timers, and so on that enhance its operation but tend to make troubleshooting more of a challenge.

Many medium-sized restaurants, hotels, and supermarkets have on-site electricians but no licensed refrigeration technicians. At these occupancies, in case of malfunction, it is necessary for the electrician to make the initial evaluation and perform troubleshooting procedures before making the decision to call in an individual who has an EPA license. This permit is not required as long as you do not break into the refrigeration circuit, which, of course, would be involved in adding refrigerant or replacing the compressor or any of the online components or working on a leak. That being said, there are many repairs beyond the initial troubleshooting procedure that can be made by the on-site or contracting electrician.

The original complaint is usually made by a kitchen or store worker, and the electrician receives it. The temperature has deviated, or there is an unusual noise.

All walk-ins and reach-ins should be equipped with a digital temperature readout on the outside wall or a standard thermometer placed inside. The thermostat should be set to maintain the temperature appropriate for the contents. Freezers are set quite low, sometimes –20°F. A lower temperature preserves the contents better and provides more time for repairs in case of malfunction. But it is more expensive to power and maintain a low temperature because the thermostat will force the compressor to work long hours. Ice cream freezers are set at +20°F so that the material is easy to scoop. Coolers should be kept some where in the +40°F neighborhood—high enough to avoid damage from freezing. Produce coolers are kept a little higher because items such as lettuce freeze very easily. Wine coolers have unique imperatives. The temperature and humid ity must be carefully controlled.

Most operator complaints fall into two categories, and they are easily resolved. If the temperature falls too low or is running just a little high, the thermostat probably needs to be adjusted. The best practice is to tweak it just the slightest amount and then to come back and check it an hour later.

If the temperature rises substantially, first check to see if the compressor is running. If it is, feel the high-pressure line on both sides of the condenser to see if it is hot. This will provide a quick overview of the system status.

When the compressor is running but the temperature in the box is high, most of the time the evaporator is in need of defrosting. This happens from time to time. An accumulation of moisture builds up on the evaporator, starting as a thin layer of frost and eventually becoming a thick coating of ice. Even though the fan blows across the frozen block, there is no significant cooling, and eventually, the temperature of the box approaches the ambient temperature outside of it.

The remedy is to get rid of the ice, and then the refrigeration equipment will function as it should, maintaining a temperature in accordance with the setting of the thermostat.

If the box is not needed right away, for example, if the contents can be transferred to another unit, the repair is very easy. Simply power down the compressor and prop open the door. A small heater or fan can be directed toward the evaporator. After a few hours, the ice will be gone, and the compressor can be restarted.

The best course of action is to take preventive measures. If the freeze-up goes unnoticed, large amounts of valuable stock can be lost. A walk-in freezer can easily hold $15,000 worth of food. Also, an accumulation of ice can block the fan, preventing the motor from turning and possibly burning up the windings.

Kitchen or grocery workers should be asked to check the temperature inside the box on a regular basis, especially during the heat of summer when the humidity is high. They should *never* leave the door open or even ajar because the source of moisture is always from outside. They should organize the work so that as many tasks as possible are performed in a single trip inside, and the door is not opened unnecessarily. Besides causing freeze-ups, an open door makes the compressor work long hours, resulting in more frequent replacements.

Another common cause of evaporator frosting, especially in small reach-ins, is a worn door gasket, so gasket material should be kept in stock and used before the wear becomes critical.

If the refrigeration unit is full of stock and must be defrosted immediately, it has to be done manually. First, power down the evaporator fan and compressor so that you will not be battling the cold. Make sure that the fan cannot restart automatically. Just to be sure, keep your hands clear of the fan blades at all times because terrible injuries are possible. Using a combination of propane torch, heat gun, and small portable heater, melt all the ice on the evaporator core. Where possible, as they loosen, remove chunks of ice. It is surprising how long it takes to defrost an evaporator that has acquired a thick coating of ice. Expect to spend over an hour. Using a propane torch, great care must be taken not to damage the aluminum fins of the evaporator or any wiring that may have become embedded in the ice.

Several methods have been developed to control freezing. A timer can be inserted in series with all ungrounded power-supply conductors, shutting down the compressor for an hour several times during each 24-hour period (Figure 6-13). Also on a timer, heating elements are sometimes embedded in the evaporator core, and they are very effective at controlling icing.

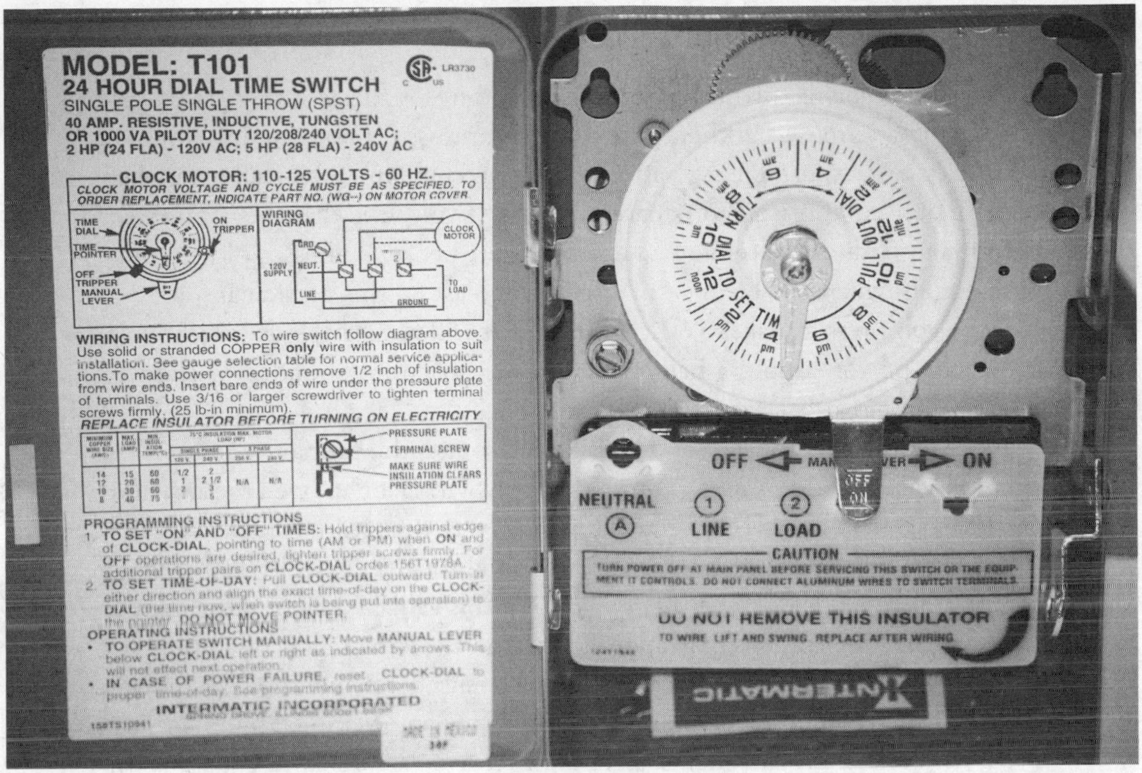

FIGURE 6-13 • A 24-hour time switch is used to control the defrost cycle for refrigeration, cycle a hot water heater, and control lighting or similar loads.

The two repairs I have mentioned, adjusting the thermostat and defrosting the evaporator, are effective perhaps 80 percent of the time. The remaining 20 percent may prove elusive and a true test of your troubleshooting skills. These are most applicable when the compressor refuses to run even when you turn the thermostat all the way down to its coldest setting. (Don't forget to restore the correct setting when you are done!)

There are a number of controls that affect operation of the motor. Temperature and pressure sensors are wired to the motor controller, and either of them can shut down the motor. Also, it is not uncommon for the timer to become defective and keep the motor from running. If you have access to a schematic, check all these control systems. Look for bad sensors, bad wiring, and bad relays. If a relay is stuck open, lightly tapping it with a wood dowel may cause it to close so that the compressor starts. Sensors can be shunted with an insulated jumper wire, allowing the compressor to start. But this method does not differentiate between a bad sensor and a good sensor that is doing its job in response to a dysfunction elsewhere.

Of course, one of the first diagnostic procedures is to see if there is voltage at the motor terminals. If there is and the compressor is not running, you will need to have a licensed refrigeration technician come in and install a new hermetic motor compressor. Work carefully here. If you state that the compressor is bad, and if the replacement unit won't start because of a simple circuitry problem elsewhere, you will be in disgrace, at least on this particular job.

If the compressor makes a 60-cycle hum but will not start, look for low voltage in the power supply, a bad start relay or capacitor, grounded or shunted motor windings, or a seized compressor.

If the motor will not start and it makes no sound, the problem can be a bad circuit breaker or motor protector, open supply wiring, or an open motor winding. Other possibilities are defective pressure or temperature controls.

An excellent diagnostic tool is the inline sight gauge. It will tell you if the refrigerant level is low. The difficulty is that if this gauge is full of refrigerant or full of air, both conditions may appear the same. If there are lots of air bubbles, more refrigerant needs to be added by a licensed refrigeration technician (Figure 6-14). The normal condition is to see only an occasional bubble.

As with all electrical systems, a good preventive-maintenance program will keep things running smoothly with no interruption of service, and there will be great savings in parts and labor.

Any motors should be lubricated if indicated on the nameplate. If the heat exchangers that are associated with the compressors (condensers) are air-cooled, the fans tend to draw in a lot of dust, which ends up clogging the cores. This makes for inadequate cooling of the hot refrigerant. The result is less efficient cooling in the box, so the compressor has to work long hours, making for more expensive operation. Condensers should be vacuumed out periodically. Also check V-belt tension, and replace belts that are worn.

The motor compressor and fan motors should be checked by means of a clamp-on ammeter to see if they are drawing increased amounts of current. Maintenance logs should be posted nearby so that trends may be spotted before they become acute.

Part of the maintenance program should be to perform an overall evaluation to see if the original installation is efficient and properly constructed and to make sure that the wiring and equipment installations are Code-compliant. To begin, you will want to see if the branch-circuit conductors are properly sized. Hermetic compressor branch circuits are figured differently from other motor

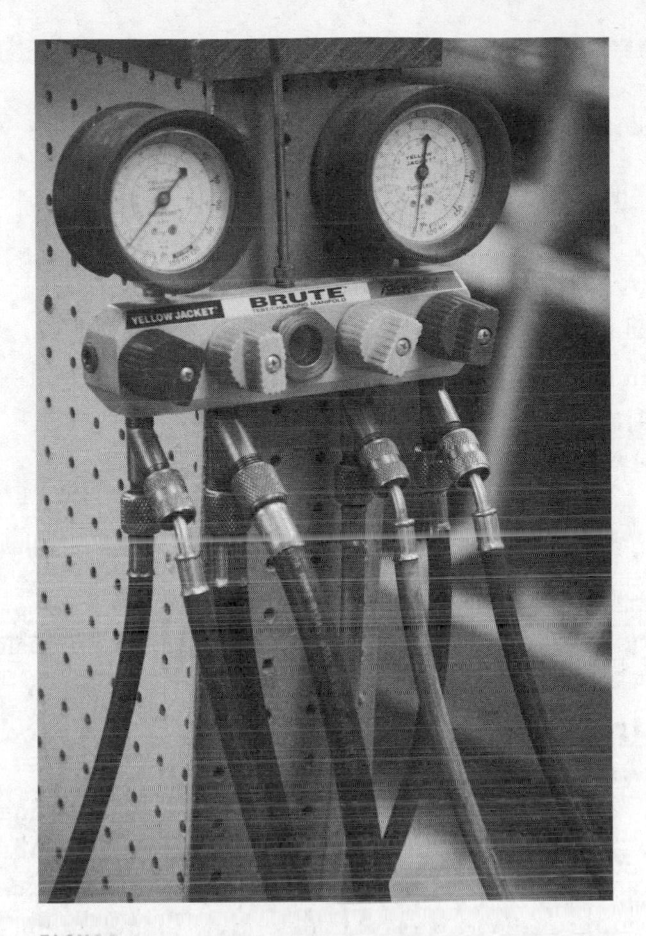

FIGURE 6-14 • Refrigerant gauge manifold set for adding or removing refrigerant without release into the atmosphere.

branch circuits. The nameplate provides the branch-circuit selection current, and you need to refer to this value when sizing the circuit. The motor compressor branch-circuit short-circuit and ground-fault protective device is to be capable of carrying the starting current of the motor. A protective device having a rating or setting not exceeding 175 percent of the motor-compressor-rated load current or branch-circuit selection current, whichever is greater, is permitted. Where the protection specified is not sufficient for the starting current of the motor, the rating or setting may be increased but is not to exceed 225 percent of the motor rated-load current or branch-circuit selection current, whichever is greater.

The conductors, then, are sized in the normal manner in accordance with the Chapter 3 ampacity tables.

Article 445 covers generators. It is brief because the internal circuitry is not under *NEC* jurisdiction. Grounding was covered previously in Article 250. Specific applications are covered in respective articles:

- Fire pumps: Article 695
- Emergency systems: Article 700
- Legally required standby systems: Article 701
- Optional standby systems: Article 702
- Interconnected electrical power-production sources: Article 705
- Critical-operations power systems: Article 708

A generator installation will involve requirements that appear in many different Code locations, so it is necessary to research carefully for any job that is contemplated. Unsettled weather has caused frequent utility outages. As a result, generator installations, along with the necessary transfer switch, have provided a great amount of work for electricians who have made this a specialty.

Inductance and the Transformer

Article 450, "Transformers and Transformer Vaults (Including Secondary Ties)," covers all transformer installations, with eight exceptions. If the contemplated transformer does not correspond to any of the exceptions, it is necessary to go through Article 450 to determine the requirements that pertain to the particular job.

Transformers range in electrical capacity and size from units that can fit in the palm of your hand or are soldered onto a printed circuit board to units that have to be set in place by means of a crane. Most transformers consist of two or more insulated windings, with or without taps placed at specific intervals within the windings. There is often an iron core. Transformers that operate at high frequencies more likely have an air core.

In its simplest form, a transformer consists of primary and secondary windings on an iron core. The primary winding has two input leads, and the secondary winding has two output leads. The voltage that is fed to the primary is stepped up or stepped down depending on the ratio of the number of turns in the two windings. If the primary has 100 turns and the secondary has 200 turns, the voltage that is measured at the output leads of the secondary winding will be double the voltage connected to the input leads of the primary winding. Such a transformer is known as a *step-up transformer*. It would be useful if you had a 120-V generator and wanted to run a 240-V load. Of course, we cannot get free energy. The actual power in both primary and secondary is the same, except for a small loss, which appears as heat that is dissipated by the transformer. This loss is measured by the efficiency rating of the transformer, which is expressed as a percentage and is often about 90 percent.

If the 240-V load that is connected to the secondary draws 2 A, the primary will draw 4 A (not figuring in transformer loss).

A *step-down transformer* is similar, but all the numbers are reversed. If the primary winding has 200 turns and the secondary has 100 turns, 240 V can be transformed to 120 V. If the load connected to the secondary is 4 A, 2 A will flow in the primary circuit. A familiar sight throughout neighborhoods that have aerial utility lines is the step-down transformer at a residential service. The high line voltage is stepped down to the 120/240-V single-phase service level. The secondary operates at 240 V with a grounded center tap that measures 120 V with respect to either of the hot legs.

Still Struggling

Voltage can be increased or decreased by varying the turns ratio in a transformer. When voltage goes up, amperage goes down. Power remains the same, except for a small loss in the transformer.

It is also possible for the primary and secondary windings to have the same number of turns. There will be no change in voltage or amperes. This device is known as an *isolation transformer*. One side of the primary may be grounded, but this grounding does not pass through to the secondary unless there is either an internal fault or an intentional external connection. The isolating transformer is used in areas of healthcare facilities where it is desired to eliminate the ground, and it is also used to suppress electrical noise in sensitive devices. Another function is to transfer power between two circuits that are not intended to be connected.

Nearly all electrical equipment that contains semiconductors will have a power supply that delivers low-voltage dc to bias these devices. At the front end of the power supply there is invariably a transformer. It is easily recognized by its laminated core and direct connection of the primary to the incoming ac line and multiple secondary taps to supply whatever voltages are needed.

It is to be emphasized that a transformer will operate only on ac. If dc is connected to the primary, there will be no output at the secondary, except for a pulse when the dc is switched on and another when it is switched off.

When all or some parts of a piece of electrical equipment appear dead, it is natural to suspect the power supply, and a frequent offender is the transformer. The secondary voltages are usually marked on the transformer, printed circuit boards, or the schematic. They can be measured with a voltmeter while the equipment is powered up. But be sure that you do not contact any high voltages. Even the insulated probe may not be adequate protection. The rule should be: Do not measure anything unless you know what to expect. Even after the chassis is powered down, power-supply capacitors may store dangerous high-voltage energy. Do not work in this area unless you know how to discharge the capacitors, and remember that even after this is done, the equipment may have distributed capacitance sufficient to store hazardous electrical energy.

As an alternative to measuring secondary voltage while the equipment is powered up, you can shut it down, discharge stored energy, and then use your ohm function to check out the windings. To do this, it is sometimes necessary to take the transformer out of the circuitry because any parallel low impedance will disrupt your readings. For each two-wire circuit, it is only necessary to cut or unplug one wire. Then measure the primary and all secondaries. The normal readings should be quite low ohms. The general rule is the smaller (physically and electrically) the transformer, the higher is the resistance. But dc resistance will always be quite low. The ac impedance will be much higher due to inductive reactance. Try wiggling the leads where they go into the transformer because this will sometimes bring out an intermittent. Also, occasionally a transformer will appear normal when it is cold, but on heating up, the fault will appear. A heat gun is useful as a diagnostic tool if this is suspected.

A bad transformer will usually have either open or shorted windings, so its condition should be obvious. However, sometimes adjacent windings will short out, changing the secondary voltages to some degree. Also, shorted primary windings can unexpectedly increase secondary voltages because the windings ratio is changed.

Chapter 4 concludes with five articles dealing with additional types of equipment for general use, such as phase converters and storage batteries and with equipment operating at over 600 V.

Hazardous Locations

Chapter 5, "Special Occupancies," covers some of the core topics that electricians confront, notably the dense and challenging subject of hazardous locations. (They are sometimes called *classified locations*. Either term is correct.)

Keeping in mind that the *NEC* is concerned with the practical safeguarding of persons and property from hazards arising from the use of electricity, it is obvious that certain areas require special attention because of the presence of fuels that ignite easily.

Hazardous locations are divided into three classes defined according to the material present. Class I areas contain flammable gases, flammable liquid-produced vapors, and combustible liquid-produced vapors. Class II locations contain combustible dusts. Class III locations contain ignitable fibers or flyings. Articles 500 through 504 cover the requirements for electronic and electrical equipment and wiring for all voltages in these three classified locations. Special methods and materials are specified for each location. Generally, Class I locations are considered the most hazardous, Class II intermediate, and Class III least dangerous. But this is not a totally valid way to look at it, and it would be no comfort to victims of a fiery inferno that occurred because of faulty wiring in a Class III location. Interestingly, certain explosion proof enclosures that are compliant for a Class I location are not adequate for a Class II location because the nature of the materials that define the location require totally different protective techniques.

Each of the three classes is subdivided into two divisions. In Division 1, the hazard is more immediate, for example, where flammable liquids are transferred from one container into another so that vapors are released into the air in the course of normal operations.

A Class I, Division 2 location is one

- In which volatile flammable gases, flammable liquid-produced vapors, or combustible liquid-produced vapors are handled, processed, or used but in which they will normally be confined within closed containers or closed systems from which they can escape only in case of accidental rupture or breakdown of such containers or systems or in case of abnormal operation of equipment, or
- In which ignitable concentrations of flammable gases, flammable liquid-produced vapors, or combustible liquid-produced vapors are normally prevented by positive mechanical ventilation and which might become hazardous through failure or abnormal operation of the ventilating equipment, or

- That is adjacent to a Class I, Division 1 location and to which ignitable concentrations of flammable gases, flammable liquid-produced vapors, or combustible liquid-produced vapors above their flash points might occasionally be communicated unless such communication is prevented by adequate positive-pressure ventilation from a source of clean air and effective safeguards against failure are provided.

The same sort of distinctions are made between Division 1 and Division 2 in each of the three classes. So you can see that there are six types of locations, or seven if you count unclassified locations, which are not as hazardous. Each of these locations has different requirements as far as wiring and equipment are concerned, and there is a range of protective techniques that may be used to mitigate the hazards, some overlapping and some unique to a single type of area.

Electricians have to provide lighting, power electrical equipment, and wire sensors in the sensitive areas, and there are some strategies for accomplishing this without risking disaster. The best approach is to design the installation so that the electrical wiring and equipment are outside the hazardous area or in a less hazardous area, such as in a Division 2 rather than a Division 1 location. A spray booth, for example, can be illuminated by means of light fixtures that are located outside the hazardous-area boundaries. This is always the preferred method for dealing with a hazardous location and should be used where possible.

If it is judged necessary to install the electrical equipment and wiring within a hazardous location, a number of protective techniques are available to mitigate the danger that the hazardous material will be ignited. These techniques may be used singly or in combination:

- Explosion-proof equipment is permitted in Class I, Division 1 or 2 areas.
- Dust ignition-proof equipment is permitted in Class II, Division 1 or 2 areas.
- Dust-tight equipment is permitted in Class II, Division 2 or Class III, Division 1 or 2 areas.
- Purged and pressurized equipment is permitted in any hazardous area for which it is identified.
- Intrinsic safety is permitted for equipment in any hazardous area.
- Nonincendive circuits, nonincendive equipment, nonincendive components, and hermetically sealed equipment are permitted for equipment in Class I, Division 2, Class II, Division 2, or Class III, Division 1 or 2 areas.
- Oil immersion is permitted for contacts in Class I, Division 2 areas.

- Combustible gas-detection systems are permitted as a means of detection only in industrial establishments with restricted public access and where conditions of maintenance and supervision ensure that only qualified persons service the installation.

Protection techniques that are permitted for a Division 1 location may be used for a Division 2 location of any given class. This is so because the hazards are the same, although not as immediate in Division 2. This interchangeability is not applicable among classes, however. Explosion-proof enclosures that provide protection in a Class I location are not suitable for a Class II location, where a coating of dust could cause a hazardous temperature rise.

Locations may be simultaneously Class I, II, and III, and it is necessary to comply with all requirements. Dual-rated equipment is available, but material and temperature ratings must be observed as well.

Additional Code requirements address wiring in hazardous areas. An important element is raceway sealing. The purpose is to ensure that hazardous gases or liquids do not migrate from Division 1 to Division 2 areas or into unclassified areas where enhanced protective techniques are not in place. Specialized conduit fittings are available to permit introduction of the sealing material where required, after the wiring has been installed. Enhanced grounding and bonding techniques are needed. Grounding lugs make redundant bonding possible, offering a further level of protection against arcing (Figure 6-15).

FIGURE 6-15 • Bushing with grounding lug used where redundant grounding is required.

Hazardous location wiring and equipment are far more costly in material and labor than similar work in unclassified areas. Designers and electricians generally look for ways to reduce costs without compromising safety. I have already mentioned the most effective technique, which is to locate the equipment and wiring in an unclassified or less hazardous location, where possible.

Another strategy is to employ ventilation. A good example is in the commercial garage. If there are pits or below-floor work areas, such as for lubrication, use of ventilation can pave the way for a less stringent classification. The area becomes unclassified if there is sufficient ventilation to provide at least four air changes per hour extending across the entire floor with exhaust air removed within 12 in. of the floor.

If this ventilation is not provided, the floor area up to 18 in. above the unventilated pit and below-grade or subfloor work area, extending horizontally 3 ft from its edge, must be considered Class I, Division 2, an area where electrical wiring and equipment are costly. Moreover, areas bordering classified locations within a commercial garage where flammable vapors are not likely to be released, such as switchboard or stock rooms, become unclassified if mechanically ventilated at a rate of at least four air changes per hour providing positive pressure or if they are isolated from the hazardous area by means of partitions.

Still another hazard mitigation technique that allows great savings in installation is the use of intrinsically safe systems. For explosion or fire to take place, three elements must come together. They are a source of ignition such as an electrical spark or an overheated surface, a supply of fuel such as a volatile liquid, and oxygen, available from the atmosphere. Intrinsically safe systems eliminate the source of ignition so that fire or explosion cannot take place. Accordingly, intrinsically safe equipment may be installed in any hazardous location for which it is listed. Because of the low power levels, it is not possible to use this type of system to power motors or industrial equipment. It is used for controls, instrumentation, and power-limited fire-alarm systems. General-purpose rather than explosion-proof enclosures may be used.

In designing an intrinsically safe system, the objective is to limit the applied voltage and connected load so that power levels will not be sufficient for ignition to take place. Then it is necessary to ensure that there is not unintended infiltration of higher voltages from outside the hazardous location due to malfunction of equipment, miswiring, or a power surge. To prevent this from happening, observe required separations between intrinsically safe apparatus and associated field wiring.

These specifications are contained within the control drawings furnished by the manufacturer. They contain permitted interconnections between intrinsically safe apparatus and associated field wiring. For a safe installation, it is essential to adhere to the control drawings.

To ensure isolation, a Zener diode is installed outside the hazardous location barrier. It is wired across the input terminals, and because of the intentional (nondestructive) reverse-bias breakdown voltage, unwanted energy is prevented from infiltrating the hazardous area. For this to work, there must be an impeccable ground connection. Two redundant ground connections are used, each with a resistance of under 1 Ω. Grounding conductors are protected from physical damage and are insulated. The ground electrode should be located as close as possible.

TIP ***Do not attempt wiring within a hazardous area unless you are thoroughly schooled in this exacting and complex work.***

Commercial Garages

Article 511, "Commercial Garages, Repair and Storage," is one of the more important *NE*C articles. A frequent Code violation involves substandard wiring in one of these facilities. The Code permits Type NM (Romex) cable to be installed in the attached or detached garage that goes with a one- or two-family dwelling. Commercial garages should be wired, at a minimum, using EMT and Type MC cable, and additional requirements are triggered for any hazardous areas within the garage.

It is common to see a repair garage built on residential property that is intended as a backyard operation. Obviously, gray areas are possible. But if there is the prospect that fuel systems will be opened or welding equipment used, you would definitely want to consider the garage commercial, which means fully recognizing any hazardous areas and installing wiring in the unclassified areas in raceways.

TIP ***Many small commercial garages are improperly wired using Type NM (Romex) cable.***

It is required also that all 125-V, single-phase, 15- and 20-A receptacles installed in areas where electrical diagnostic equipment, electrical hand tools, or portable lighting equipment is to be used must have GFCI protection. This would include virtually the entire garage with the exception of an office space.

This requirement is particularly important in a commercial garage because the floor is usually a slab in contact with the earth, and there is the potential, at times, of standing water. Hand tools and trouble lights may have frayed cords and are frequently run to failure.

Healthcare Facilities Including Hospitals

Article 517, "Healthcare Facilities," contains numerous exacting requirements for very good reason. Patients, already in a weakened state, may have a conductive metal instrument placed in the bloodstream so that there is a low-impedance connection to the heart. A very low-level voltage can disrupt the tenuous cardiac rhythm. There are numerous other issues as well. Patients on life support require continuity of the power supply to sustain life. Moreover, any electrical outage, partial or facility-wide, will negatively affect the quality of the healthcare.

In a healthcare facility, the systems and subsystems are highly integrated. Fire alarms, elevators, telephone, ventilation, and other components interact with each other and with the facility's electrical system. To design and build an infrastructure of this magnitude, especially the electrical portion, requires a great amount of knowledge and expertise. For a completed, functioning healthcare facility, the maintenance is ongoing and exacting. At all times, the electricians have to understand Code mandates for healthcare facilities so that their work will be in compliance.

Besides the *NEC*, two other documents contain requirements for healthcare facilities, and if you plan to design and install wiring in these occupancies, you will need to become familiar with them. They are National Fire Protection Agency (NFPA) 99-2005, *Standard for Healthcare Facilities*, and NFPA 101-2009, *Life Safety Code*. These volumes are referenced in *NEC 2011*, and all three documents work together to determine the materials and procedures for wiring healthcare facilities.

To get a sense of what is involved, consider these *NEC* definitions:

- *Ambulatory healthcare occupancy.* A building or portion thereof used to provide services or treatment simultaneously to four or more patients that provides, on an outpatient basis, one or more of the following: (1) treatment that renders the patients incapable of taking action for self-preservation under emergency conditions without the assistance of others, (2) anesthesia that renders patients incapable of taking action for self-preservation under emergency conditions without assistance of others, and (3) emergency or urgent care for patients who, due to the nature of their injury or illness, are incapable of taking action for self-preservation under emergency conditions without the assistance of others.

- *Anesthetizing location.* Any area of a facility where a flammable or non-flammable inhalation anesthetic may be used. This includes their use for relative anesthesia, where the patient is not made entirely unconscious. Flammable inhalation anesthetics are no longer generally used in the United States, but the Code still refers to them.
- *Critical branch.* A subsystem of the emergency system that consists of feeders and branch circuits supplying electricity for lighting, special power circuits, and selected receptacles serving areas and functions related to patient care and that are connected to alternate power sources by one or more transfer switches during interruption of the normal power source.
- *Alternate power source.* A motor-driven generator(s) that will provide power during a utility outage. Sometimes backup batteries perform this function.
- *Electrical life-support equipment.* Electrically powered equipment whose continuous operation is necessary to maintain a patient's life.
- *Emergency system.* A system of circuits and equipment intended to supply alternate power to a limited number of prescribed functions vital to the protection of life and safety.
- *Equipment system.* A system of circuits and equipment arranged for delayed, automatic or manual connection to the alternate power source and that serves primarily three-phase power equipment.
- *Essential electrical system.* A system comprised of alternate sources of power and all connected distribution systems and ancillary equipment designed to ensure continuity of electrical power to designated areas and functions of a healthcare facility during disruption of normal power sources and also to minimize disruption within the internal wiring system.
- *Exposed conductive surfaces.* Surfaces that are capable of carrying electric current and that are unprotected, unenclosed, or unguarded, permitting personal contact. Paint and anodizing and similar coatings are not considered suitable insulation.
- *Hazard current.* For a given set of connections in an isolated power system, the total current that would flow through a low-impedance object if it

were connected between either isolated connector and ground. The fault hazard current is the hazard current of a given isolated system with all devices connected except the line-isolation monitor. The monitor hazard current is the hazard current of the line-isolation monitor alone. The total hazard current is the hazard current of a given isolated system with all devices, including the line-isolation monitor, connected.

- *Healthcare facilities.* Building or portions of buildings in which medical, dental, psychiatric, nursing, obstetrical, or surgical care is provided. They include but are not limited to hospitals, nursing homes, limited-care facilities, clinics, medical and dental offices, and ambulatory-care centers, whether permanent or movable.
- *Hospital.* A building or portion thereof used for 24-hour medical, psychiatric, obstetric, or surgical care for four or more inpatients.
- *Isolated power system.* A system comprising an isolating transformer or its equivalent, a line-isolation monitor, and its ungrounded circuit conductors.
- *Isolation transformer.* A transformer of the multiple-winding type with the primary and secondary windings physically separated that inductively couples its secondary winding(s) to circuit conductors connected to the primary winding(s).
- *Life-safety branch.* A subsystem of the emergency system consisting of feeders and branch circuits intended to provide adequate power needs to ensure safety to patients and personnel and that are automatically connected to alternate power sources during interruption of the normal power source.
- *Limited-care facility.* A building or portion thereof used for 24-hour housing of four or more persons who are incapable of self-preservation because of age, physical limitation due to accident or illness, or limitations such as mental retardation/developmental disability, mental illness, or chemical dependency.
- *Line-isolation monitor.* A test instrument designed to continually check the balanced and unbalanced impedance from each line of an isolated circuit to ground and equipped with a built-in test circuit to exercise the alarm without adding to the leakage-current hazard.
- *Nurses' stations.* Areas intended to provide a center of nursing activity for a group of nurses serving bed patients, where the patient calls are received, nurses are dispatched, nurses' notes are written, inpatient charts are prepared, and medications are prepared for distribution to patients.

- *Nursing home.* A building or portion of a building used for 24-hour housing and nursing care of four or more persons who, because of physical or mental incapacity, might be unable to provide for their own needs and safety without the assistance of others.
- *Patient-bed location.* The location of a patient sleeping bed or the bed or procedure table of a critical-care area.
- *Patient-care area.* Any portion of a healthcare facility wherein patients are intended to be examined or treated. Areas of a healthcare facility in which patient care is administered are classified as general-care areas or critical-care areas. Business offices, corridors, lounges, day rooms, dining rooms, and similar areas are not patient-care areas.
- *General-care areas.* Patient bedrooms, examining rooms, treatment rooms, clinics, and similar areas in which it is intended that the patient will come in contact with ordinary appliances such as a nurse call system, electric beds, examining lamps, telephones, and entertainment devices.
- *Critical-care areas.* Special-care units, intensive-care units, coronary-care units, angiography laboratories, cardiac catheterization laboratories, delivery rooms, operating rooms, and similar areas in which patients are intended to be subjected to invasive procedures and connected to line-operated electromedical devices.
- *Wet-procedure locations.* Spaces within patient-care areas where a procedure is performed and that are normally subject to wet conditions while patients are present. These include standing fluids on the floor or drenching of the work area, either of which procedure is intimate to the patient or staff. Routine housekeeping procedures and spillage of liquids do not define a wet-procedure location, nor do bathrooms, lavatories, or sink areas.
- *Patient-care vicinity.* In an area in which patients are normally cared for, the patient-care vicinity is the space with surfaces likely to be contacted by the patient or an attendant who can touch the patient. Typically, in a patient room, this encloses a space within the room not less than 6 ft beyond the bed in its nominal location and extending not less than 7½ ft above the floor.
- *Patient equipment-grounding point.* A jack or terminal that serves as the collection point for redundant grounding of electrical appliances serving a patient-care vicinity or for grounding other items in order to eliminate electromagnetic interference problems.

- *Psychiatric hospital.* A building used exclusively for 24-hour psychiatric care of four or more patients.
- *Reference grounding point.* The ground bus of the panelboard or isolated power system panel supplying a patient-care area.
- *Selected receptacles.* A minimum number of electrical receptacles to accommodate appliances ordinarily required for local tasks or likely to be used in patient-care emergencies.

By going through the preceding definitions, you can get a good overview of the way in which the *NEC* organizes healthcare facility wiring. Points to watch are different types of healthcare facilities, grounding rules, and alternate power source and loading geometry.

It is to be emphasized that healthcare facilities, as covered in Article 517, do not include veterinary offices or animal clinics. They do, however, include areas of a larger building such as a clinic within an office complex.

PROBLEM 6-3

Why is the alternate power source important in a healthcare facility?

SOLUTION

The lives of some patients depend on having continuous power for life-support equipment. Moreover, a healthcare facility could not function efficiently during a total outage.

Part II of Article 517 deals with wiring and protection in healthcare facilities. It does not apply to areas in nursing homes and limited-care facilities that are designated only as patient sleeping areas. Also, it is not applicable to business offices, corridors, and waiting rooms in outpatient facilities, dental and medical offices, and clinics.

Grounding in Healthcare Facilities

Section 517.13, "Grounding of Receptacles and Fixed Electrical Equipment in Patient-Care Areas," focuses on the most important protective measure that prevents electric shock of patients. It specifies that all branch circuits that serve patient-care areas must have an effective ground-fault current path back to the panel. This is achieved by installing the branch circuit in a metal raceway system or a cable having a metallic armor or sheath assembly that qualifies as an equipment-grounding conductor. This is in addition to the green insulated equipment-grounding conductor that is within the raceway or cable. This arrangement is known as *redundant grounding*, and the purpose is to ensure that the grounding path is not broken, an event that would expose patients to a risk of electric shock. If armored cable is used, it must be Type AC, which has a metallic strip on the inside of the armor in intimate contact with it in order to enhance the grounding continuity. The requirement applies to patient-care areas in hospitals, nursing homes, clinics, and medical and dental offices. Besides patient rooms, the redundant grounding requirement applies to recreational and therapy areas and examining rooms, to mention some examples.

The insulated grounding conductor is to attach to the grounding terminals of all receptacles, metal boxes, and enclosures containing receptacles and all non-current-carrying conductive surfaces of fixed electrical equipment that may become energized and are subject to personal contact and operating at over 100 V. The grounding conductor is to be an insulated wire, solid or stranded. It is to be installed with the branch-circuit conductors in one of the wiring methods that qualifies as an equipment-grounding conductor. The redundant ground wire does not have to go all the way back to the entrance panel, just to the nearest upstream load center that has an overcurrent device. It does not have to connect to metal faceplates, to light fixtures at least 7½ ft above the floor, or to switches located outside the patient care facility.

Another important set of requirements in healthcare facility wiring is contained in Section 517.14, "Panelboard Bonding." It provides that all equipment-grounding terminal buses of both normal and essential branch-circuit panelboards serving the same individual patient care vicinity are to be connected together by means of an insulated copper conductor at least 10 AWG. Where two or more panelboards serving the same individual patient-care vicinity are served from separate transfer switches on the emergency system, the equipment grounding terminal buses of these panelboards are to be connected together by a conductor having the same specifications. This conductor may be broken in order to terminate on the equipment-grounding terminal bus in each panelboard.

Section 517.18, "General-Care Areas," provides that each bed location is to be supplied by at least two branch circuits, one from the emergency system and one from the normal system. All branch circuits from the normal system must originate from the same panelboard. The purpose of these requirements is to ensure that there will be continuity of power for life support, monitoring, diagnostic, and therapeutic equipment and that in the event of a fault that affects a branch circuit originating at one of the panels, there will not be a dangerous voltage differential between equipment-grounding conductors. Another requirement is that for branch circuits serving patient bed locations, multiwire branch circuits are prohibited.

These requirements are not applicable to branch circuits serving only special-purpose outlets and receptacles, such as portable x-ray outlets, and patient bed locations in clinics, medical and dental offices, outpatient facilities, psychiatric and substance abuse and rehabilitation hospitals, and sleeping rooms of nursing homes and limited-care facilities.

Each patient bed location must be provided with a minimum of four receptacles. They may be single, duplex, quad, or any combination. All receptacles, even if there are more than the required four, must be listed and identified as hospital grade, having a green dot on the face of the receptacle.

Section 517.19, "Critical-Care Areas," resembles the preceding section on general-care areas, but there are enhanced requirements. Here it is especially critical to observe electrical circuit, receptacle, and grounding and bonding requirements. From the definitions, we know that critical-care areas are where patients are subjected to invasive procedures and where they are connected to line-operated electromedical equipment. One example of a critical-care area is the intensive-care unit, but there are other somewhat less demanding locations as well.

Each bed location within a critical-care area is to be served by a minimum of two branch circuits. One or more is to be connected to the emergency system, and one or more is to be connected to the normal system. A minimum of one branch circuit from the emergency system is to supply one or more outlets only at that bed location. All normal-system branch circuits are to be connected to a single panelboard. All emergency-system receptacles must be identified and marked to indicate the panelboard from which they originate and the circuit number. As in the general-care patient bed locations, multiwire branch circuits are prohibited.

If the critical-care patient bed locations are supplied through two separate transfer switches, they are not required to have circuits connected to the normal system. Receptacles connected to the emergency system are customarily color-coded red. Also, the receptacle must have the green dot showing that it is hospital grade.

Each critical-care patient bed location must have at least six receptacles, two more than required for the general-care patient bed locations. Again, they may be single, duplex, quad, or any combination.

The metal raceway that qualifies as an equipment-grounding conductor is to attach to enclosures and equipment such as panelboards and switchboards so as to provide ground continuity. One of the following bonding means must be used at each junction point or termination:

- A grounding bushing and a continuous copper bonding jumper connected to the junction enclosure or the ground bus of the panel
- Connection of feeder raceways to threaded hubs or bosses on terminating enclosures
- Other approved devices such as bonding-type locknuts or bushings

Critical-care areas may have other protective techniques that are not actually required. A common protective measure is the isolated power system, frequently used in conjunction with operating rooms. The secondary of an isolation transformer is connected to the branch circuit that conveys power for loads within the sensitive area. There are two conductors, both isolated from ground The insulation of one of these conductors is orange with a stripe having a color other than white, green, or gray. The other conductor is brown with a different stripe, also other than white, green or gray.

The isolation transformer must not be located within a hazardous anesthetizing location. The secondary conductors that run into such a location must be installed in accordance with Class I hazardous location specifications.

The system has standard overcurrent protection plus a line isolation monitor. This instrument, usually located at the nurses' station, has a green signal light that indicates that both lines are isolated from ground. Unlike in most wiring, the equipment-grounding conductor may be run outside the raceway in order to limit conductance between the two lines and ground.

Part III, "Essential Electrical System," focuses on the portion of a healthcare facility's electrical infrastructure that is necessary for life safety and orderly cessation of procedures in the event that the normal system experiences an outage.

A healthcare facility's electrical structure is made up of two parts, the essential electrical system and the nonessential electrical system. The essential electrical system receives power from an alternate source when the normal power source is interrupted.

Section 517.30 further defines the essential electrical system. It provides that the essential electrical systems for hospitals are to be composed of two separate systems, each capable of supplying a limited amount of power and lighting necessary for life safety and effective hospital operation during the time the normal electrical service is disrupted. These two systems are known as the *emergency system* and the *equipment system*.

The emergency system is further subdivided into the *critical branch* and the *life-safety branch*. The equipment system is not further subdivided. It supplies mostly three-phase and motor-driven loads, large units of electrical equipment that are necessary for patient care and hospital operation.

The only source of supply for nonessential loads is the normal source. If it cuts out, the nonessential loads are powered down and remain dormant until the normal source is restored. There is no transfer switch involved. When the normal power source is online, the essential electrical system is connected to it through one or more automatic transfer switches.

If the essential system maximum demand of a hospital is not over 150 kW, the system may be switched from one source to the other by means of a single transfer switch. If the load is larger, three automatic transfer switches are required. There will be one transfer switch for the critical branch, one for the life-safety branch, and one for the equipment system. To minimize heavy current inrush at startup, the equipment-system transfer switch incorporates an intentional time delay.

Optional loads that are not required by the Code are not to be transferred if that would overload the generating equipment. If the generator becomes overloaded, these optional loads are to be automatically shed. The premise in all this is that the generator will have less capacity than the utility service. If the generator had unlimited capacity, there would be no need to divide the facility's load into essential and nonessential systems.

The life-safety branch and the critical branch are to be kept separate from other wiring and equipment and must not occupy the same raceways, boxes, or cabinets with each other. There are some exceptions. They can coexist if in transfer-switch enclosures, if in exit or emergency luminaires supplied from two sources, or if in a common junction box attached to exit or emergency luminaires supplied from two sources.

Rules for the equipment system, as one might expect, are more lenient. This wiring may coexist with other wiring within raceways and enclosures that are not part of the emergency system.

To review, the essential electrical system is made up of the equipment system and the emergency system, which is further subdivided into the critical branch and the life-safety branch. The two parts of the emergency system are to be connected to the alternate power source within 10 seconds of interruption of normal power. Simultaneously, the automatic transfer switch disconnects them from the source of normal power so as not to endanger utility workers.

The life-safety branch is to supply power to the following loads and no other loads:

- Illumination of means of egress, that is, corridors, passageways, stairways and landings at exit doors, and all necessary ways of approach to exits
- Exit signs
- Alarm and alerting systems, including fire alarms and alarms required for systems used for the piping of nonflammable medical gases
- Communications systems
- At generator and transfer-switch locations, battery chargers for battery-powered lighting units, and receptacles at the generator and transfer switch
- Generator-set accessories to facilitate operations
- Elevator cab lighting, control, communication, and signal systems
- Automatic doors used for building egress

The critical branch of the emergency system is to provide power for lighting, fixed equipment, receptacles, and power circuits in these areas and functions related to patient care.

- Critical areas that use anesthetizing gases
- Isolated power systems
- Lighting and receptacles in patient-care areas for infant nurseries, medication preparation areas, pharmacy dispensing areas, acute nursing areas, psychiatric bed areas, ward treatment rooms, and nurses' stations
- Specialized patient-care lighting and receptacles
- Nurse call systems
- Blood, bone, and tissue banks
- Telephone equipment rooms and closets
- Lighting, receptacles, and power for general-care beds, angiographic labs, cardiac catheterization labs, coronary-care units, hemodialysis rooms, emergency rooms, physiology labs, intensive-care units, and postoperative recovery rooms
- Additional lighting, receptacles, and power as needed

The critical branch may be further subdivided.

The equipment system is to connect the following equipment to the alternate source after a suitable delay:

- Central suction systems for surgical and medical functions (These could alternately be on the critical branch.)
- Sump pumps
- Compressed-air systems, alternatively permitted on the critical branch
- Smoke control and stair pressurization
- Kitchen hood ventilation and exhaust
- Ventilation for airborne infectious/isolation rooms, protective-environment rooms, exhaust fans for laboratory fume hoods, nuclear medicine areas where radioactive material is used, and ethylene oxide and anesthetic evacuation (These may be placed on the critical branch if delayed automatic connection is objectionable.)
- Supply, return, and exhaust ventilating systems for operating and delivery rooms

As you have seen, the essential electrical system of any healthcare facility must have at least two sources of independent power. The normal source powers the entire facility. If it is interrupted, the alternate source takes over and powers only the essential system. The alternate source may be one or more generators located on the premises and powered by a prime mover, usually a diesel engine. Another setup could be a second generating unit where the normal source is a generator located onsite. Or the alternate source can be an external utility when the normal source is a generating unit located on the premises. A final configuration is a battery system located on the premises.

Nursing homes and limited-care facilities with over 150 kW or more rated essential electrical systems, because they are not required to have equipment systems, must have only two transfer switches rather than the three required for hospitals.

I have provided a general overview of healthcare facility wiring. A full treatment would require at the least a hefty volume. The work is complex and exacting. To participate in new healthcare facility electrical construction in a decision-making capacity requires knowledge, expertise, and great attention to detail. As for maintenance and retrofit, the demand for this work is increasing as the many facilities in existence continue to age. The best entry into the field, besides studying print and online resources, is to be employed by a firm that specializes in this type of construction so that you can work your way into the field in an incremental way.

The healthcare facility is an example of the special occupancies that are covered in Chapter 5. Most of them are less exacting and the requirements are less extensive than for a hospital. Nevertheless, there are critical mandates that must not be neglected. It is a good idea to review the contents of this chapter so that when a question arises either on a licensing exam or in the field, you will know where to look. Most of the articles are fairly brief, and precise answers are easy to find.

TIP ***Healthcare facility wiring is for professionals who have made a thorough study of the field. There is no room for error.***

Special attention should be focused on Article 590, "Temporary Installations." Most new construction involves the installation of a temporary service and outlets so that workers will have the use of power and lighting for building purposes. The wiring is not required to conform to all standards of a permanent installation, but certain requirements must be satisfied. Of particular importance is GFCI protection for workers. They have occasion to deal with standing water and other adverse conditions. Electrical safety must not be compromised, and it is the job of the electrician to make sure that electrical hazards are not present in the workplace. Article 590 provides guidance, and it merits a close reading.

Chapter 6, "Special Equipment," is an extension of Chapter 4. Twenty-five articles, each dealing with a special type of equipment, contain requirements for such diverse items as elevators, electric welders, pipe organs, solar and wind electrical systems, and fire pumps.

I will not discuss these categories individually. When questions arise, it is only necessary to consult the relevant article. Most of them are brief, and information is easy to access. By way of preparation, go through the contents and learn the list of topics.

Special Conditions

Chapter 7, "Special Conditions," contains 10 articles that may be considered outside the conventional electrical body of knowledge. The topics covered include

- Article 700: "Emergency Systems"
- Article 701: "Legally Required Standby Systems"
- Article 702: "Optional Standby Systems"

- Article 705: "Interconnected Electric Power Production Sources"
- Article 708: "Critical Operations Power Systems (COPS)"
- Article 720: "Circuits and Equipment Operating at less than 50 Volts"
- Article 725: "Class 1, Class 2, and Class 3 Remote Control and Signaling, and Power-Limited Circuits"
- Article 727: "Instrumentation Tray Cable, Type ITC"
- Article 760: "Fire Alarm Systems"
- Article 770: "Optical-Fiber Cables and Raceways"

Most of the topics can be easily assimilated by reading the relevant articles, so I won't go through them individually. However, two of the articles have presented difficulties for electricians struggling to become familiar with them, so I will go over them in some detail. These two topics appear in *NEC* Articles 725 and 760.

Remote-Control, Signaling and Power-Limited Circuits

Article 725 covers the infamous Class 1 through 3 wiring. (They are not to be confused with Class I through III hazardous locations.)

Article 725 is sometimes mistakenly construed to be about low-voltage wiring. This is not a good way to look at it because it misses the main point. Moreover, low voltage, rather than being precisely defined, varies with the context. It is taken to mean less than 2,000, 1,000, 600, 50, and 12 V. None of these cutoff points defines the three classes of wiring discussed in Article 725.

A further source of confusion in understanding this article is that three classes are not members of the same group. Classes 2 and 3 are organized in terms of power and voltage levels, whereas Class 1 has to do with safety.

Section 725.1, "Scope," states that Article 725 covers remote-control, signaling, and power-limited circuits that are not an integral part of a device or appliance. An Informational Note elaborates, saying that these circuits are characterized by usage and electrical power limitations that differentiate them from electric light and power circuits; therefore, alternative requirements to those of Chapters 1 through 4 are given with regard to minimum wire sizes, ampacity adjustment and correction factors, overcurrent protection, insulation requirements, and wiring methods and materials.

The methods presented in Article 725 are optional alternatives to the ordinary wiring methods of Chapters 1 through 4. By far the most important of the classes for electricians is Class 2. As with the other classes, power and voltage levels are applicable, and they are given in Code tables. But it is not necessary to consult these tables, for the most part, because the power source, usually a transformer, will be listed and marked for Class 2 operation.

Section 725.2, "Definitions," lists terms that are applicable to this article:

- *Abandoned Class 2, Class 3, and PLTC cable.* Installed cables that are not terminated at equipment and not identified for future use with a tag.
- *Circuit-integrity (CI) cable.* Cable used for remote-control, signaling, or power-limited systems that supplies critical circuits to ensure survivability for continued circuit operation for a specified time under fire conditions.
- *Class 1 circuit.* The portion of the wiring system between the load side of the overcurrent device or power-limited supply and the connected equipment. An Informational Note refers to Section 725.41 for voltage and power limitations of Class 1 circuits.
- *Class 2 circuit.* The portion of the wiring system between the load side of a Class 2 power source and the connected equipment. Owing to its power limitations, a Class 2 circuit considers safety from a fire-initiation standpoint and provides acceptable protection from electric shock.
- *Class 3 circuit.* The portion of the wiring system between the load side of a Class 3 power source and the connected equipment. Owing to its power limitations, a Class 3 circuit considers safety from a fire-initiation standpoint. Because higher levels of voltage and current than for Class 2 are permitted, additional safeguards are specified to provide protection from an electric shock hazard that could be encountered.

Parts II and III define these classes in greater detail.

Section 725.31, "Access to Electrical Equipment behind Panels Designed to Allow Access," prohibits an accumulation of wires and cables that prevents removal of panels, particularly suspended ceiling panels. Cables cannot be allowed to lie directly on the panels. They are to be hung from framing or attached directly to the ceiling surface above.

Section 725.24, "Mechanical Execution of Work," provides that Class 1, Class 2, and Class 3 circuits are to be installed in a neat and workmanlike manner. Cables and conductors installed exposed on the surfaces of ceilings and sidewalls are to be supported by the building structure in such a manner that the cable will not be damaged by normal building use. The cables are to be supported by straps, staples, hangers, cable ties, or similar fittings designed and installed so as not to damage the cable.

Section 725.25, "Abandoned Cables," provides that the accessible portions of abandoned Class 2, Class 3, and PLTC cables are to be removed. The purpose of this requirement is to avoid large accumulations of abandoned cable that would provide fuel in the event of fire, resulting in a large quantity of toxic smoke. The requirement to remove abandoned cable with substantially the same wording also applies to communication and data wiring and is repeated in Articles 640, 645, 760, 770, 800, 820, and 830. Strangely, abandoned power and light cable is not required to be removed.

Section 725.30, "Class 1, Class 2, and Class 3 Circuit Identification," states that these circuits are to be identified at terminal and junction locations in a manner that prevents unintentional interference with other circuits during testing and servicing.

Section 725.31, "Safety Control Equipment," covers Class 1 circuits. Remote-control circuits for safety-control equipment are to be classified as Class 1 if failure of the equipment to operate introduces a direct fire or life hazard.

A nurse call system is not a Class 1 circuit because such a system does not initiate hazards but merely reports them. However, a nurse call system could become Class 1 due to voltage and power levels that would preclude Class 2 or Class 3 designation. Similarly, water temperature–regulating devices and room thermostats are not considered Class 1. What is confusing sometimes to novice electricians is that two differing and unrelated conditions may trigger inclusion in Class 1: higher voltage or power levels and the fact that damage to remote-control circuits of safety-control equipment would introduce a hazard. All conductors of the remote-control circuits are to be installed in rigid metal conduit, intermediate metal conduit, rigid nonmetallic conduit, electrical metallic tubing, Type MI cable, or Type MC cable or otherwise be suitably protected from physical damage.

The interesting thing about Class 1 through 3 circuits is that some have requirements that are relaxed with respect to Chapters 1 through 3 wiring methods, whereas others have more stringent requirements.

Class 1 circuits are to comply with Parts I and II of Article 725. Class 2 and Class 3 circuits are to comply with Parts I and III of Article 725. Understanding and properly applying Article 725 consists of two operations: determining the correct classification for any given circuit you intend to install and ascertaining the correct wiring methods and materials to be used. The most frequent application of Article 725 is for Class 2 wiring. An example of Class 2 wiring is the control circuit for a residential oil furnace that is fed through a wall-mounted room thermostat (Figure 6-16).

FIGURE 6-16 • A wall-mounted thermostat, typically used to control a residential oil furnace, may be wired by means of a Class 2 circuit.

The general Code rule is that the minimum size conductor for any circuit is 14 AWG copper. But Class 2 wiring may be 16 or 18 AWG depending on the maximum current that the load will draw. A control circuit may need to activate only a magnetic relay or its electronic equivalent, and that requires a small amount of current.

Still Struggling

Article 725 may seem baffling at first, but after some experience wiring Class 2 circuits, it will all come together.

What Is a Supervised Fire Alarm System?

The other challenging article in Chapter 7 is Article 760, "Fire Alarm Systems." In a commercial or industrial setting, a fire alarm system generally is integrated with elevators, sprinklers, telephone system, liquid propane gas (LPG) or natural gas supply, ventilation, fire doors, and so on. The fire alarm system is totally different from the residential-type smoke alarms powered by 9-V batteries, even if connected to the ac electrical supply and wired together to go off in concert.

Two things characterize the fire alarm system that I am discussing here. First, it is vastly more expensive than residential smoke alarms. And second, the heads and horns (or alternatives) are connected to a central control panel that has amazing functionality and includes a user interface with alphanumeric display, permitting programming, troubleshooting, and user interaction for the purpose of testing and in response to fire or false alarm.

For an electrician who works in a commercial or industrial facility, there will inevitably come a time when it is necessary to perform maintenance or repair on a fire alarm system. In order to do this, it is necessary to have a comprehensive knowledge of the system, including zone wiring methods and control-panel operation. While you may not yet have the kind of detailed knowledge that a professional fire alarm technician would possess, a certain degree of familiarity is necessary. In case of system trouble (this is a specialized term that I will discuss later in this chapter), the on-site electrician will receive the initial call. If steps cannot be taken to rectify the situation, the decision to call in a specialized fire alarm technician will be in order.

Complete coverage of this intricate topic would require, at the least, a full-length book. Here I will endeavor to provide an overview that will allow you to start doing simple procedures and put you on track to acquire more advanced fire alarm expertise. To start, it is necessary to become familiar with the following regulatory documents that pertain to fire alarm systems:

- NFPA 101, *Life Safety Code*, denotes which occupancies are required to have fire alarm systems.
- NFPA 72, *National Fire Alarm Code*, lays out overall system design parameters, such as location and spacing of heads and pull stations, testing and maintenance procedures, minimum performance requirements, and operational protocols.
- NFPA 70, *National Electrical Code*, Article 760, covers the equipment and wiring of the fire alarm system, both power to the control console and zone wiring to initiating devices and to indicating appliances and any phone lines for automatic calling. Also included are other fire alarm functions, such as guard's tour, sprinkler water flow, sprinkler supervisory equipment, elevator capture and shutdown, door release, smoke doors and

damper control, fire doors, and fan shutdown—only where the functions are actually controlled by the fire alarm system. Article 725, "Class 1, Class 2, and Class 3 Remote Control, Signaling, and Power-Limited Circuits," covers wiring emanating from the control panel. Where the circuits are power-limited, alternative requirements take effect for minimum wire sizes, derating factors, overcurrent protection, insulation requirements, and wiring methods and materials.

- Underwriters Laboratories or other inspecting agencies list all components such as control panel, smoke-detecting heads, horns, pull stations, batteries, end-of-line resistors, and any other equipment.

The primary function of a fire alarm system is to notify the occupants when combustible material within the building ignites. It would be totally unacceptable for a fire alarm system to fail to go into an alarm state in the event of fire without first going into the trouble state to indicate that there is a problem. To put it differently, a fire alarm system is far more likely to produce a false alarm than it is to fail to sound the alarm when there is a fire.

False alarms, while preferable to no alarm at all, are quite harmful. In a hotel, restaurant, or public place, at the least they are annoying to guests, and in an industrial or office setting, they interrupt the workflow and divert maintenance resources from needed tasks. And worse, repeated false alarms give rise to a complacent attitude in which the real thing may be ignored with devastating consequences. Therefore, one of the primary jobs of electricians and fire alarm technicians is to determine the causes of false alarms and work proactively to eliminate them.

Fire alarm systems most of the time are in one of three states, indicated at the user interface: normal, alarm, or trouble. When the system is normal, it means that there is no evidence of fire within the facility, and the system is ready and able to respond if there is a change. The alarm state is entered when there is evidence of fire within the area that is covered. When the system goes into alarm, loud horns and strobe lights for the hearing impaired will activate so that occupants will evacuate, maintenance workers can investigate and, if possible, extinguish the fire, and the local fire department and monitoring agency will be alerted. (Rather than loud horns, some occupancies such as nursing homes and hospitals may have softer chimes so as not to stress the patients.)

The trouble state is quite useful. Typically, a buzzer will sound. It is audible only in the vicinity of the control panel. Without disrupting operation of the entire facility, it alerts maintenance workers that there is a problem in the system that could impair proper functionality and prevent the alarm from sounding when needed.

To see how this works, you have to look at the overall structure of the fire alarm system. It is made up of three parts: control panel, initiating devices, and indicating appliances. The control panel has to be seen to be appreciated. In many stores and public places, it is located just inside the main entrance, so you can compare units made by various manufacturers. There is usually a glass door through which the user-interface alphanumeric display is visible. When the door (which may be locked) is opened, access is gained to the touchpad controls. At the bottom of the control panel or in a separate enclosure are batteries for the backup electrical supply so that, in the event of an outage or until the alternate power supply kicks in, there will be continuity of system operation. Besides the power supply, digital display, touchpad controls, and charger, the control panel contains a number of cards or printed circuit boards. Each card corresponds to one of the zones and is easily replaceable.

The second major component is the system of initiating devices. These are smoke- and heat-detecting heads, pull stations, and other types of sensors designed to detect evidence of a fire at an early stage. These initiating devices are organized into zones. In a small building, each zone could correspond to a single floor, or in a larger building, there can be two or more zones per floor. The initiating devices are wired in parallel, daisy-chained at intervals the length of the zone just like receptacles in a branch circuit. Each indicating device has four terminals, two for input and two for output to the next device.

When there is no evidence of fire, the initiating devices do not conduct. When they detect smoke or heat, they become conductive, and this is what the control panel sees. A pull station is a simple single-pole switch that shunts the lines, and this is how all initiating devices operate.

The zones are wired by means of two 16 AWG (permitted, for a power-limited fire alarm system because the zone consists of a Class 2 signaling circuit) conductors emanating from the control panel. They have a 24-V dc potential. One conductor is positive, and the other is negative. It is necessary to observe the correct polarity because the devices, containing solid-state components, have to be biased correctly. Both conductors are isolated from ground, typically run in EMT, which is grounded, as required by Article 250, because it is bonded to the control panel, which is bonded to the electrical grounding system that is part of the building's electrical service. The initiating-device conductors are made up into fire alarm cable, which has a distinctive red jacket. It is round and stiff, so it is easy to push through most raceway runs without a pull rope.

The essence of a fire alarm system is that the initiating devices, circuit integrity, and zone card are supervised at all times. So are the indicating appliance zones and the links to the sprinkler system zones. This does not involve a human sitting at a monitor watching for a malfunction. In reality, the supervisory function is made possible by the 24 V dc that originates in and is monitored by the control panel. The control panel is looking at that voltage at all times, and variations in the current flow determine whether the system will remain in the normal state, enter the alarm state, or enter the trouble state. Another function of the dc voltage is to supply bias for the solid-state components, including the LEDs that are in the heads.

How, you might ask, can the control panel detect a break in the line as opposed to the normal state when the initiating devices are not conducting? The control panel can differentiate between these two conditions by means of an ingenious arrangement that involves the end-of-line resistor. This kilo-ohm-range device is wired across the zone lines after the last initiating device. Measuring current flow, the control panel will see this resistance as the only load in the zone when there is no fault and the initiating devices are not conducting; that is, there is no evidence of fire.

If one of the two conductors becomes faulted to ground because it chafes on the inside of the metal raceway or is pinched in a connector, this will trigger a trouble state and appear as an error message on the alphanumeric display. All unaffected zones will remain functional and be capable of putting the system into the alarm state if fire is detected.

When there is a trouble state, especially one of these ground faults, maintenance workers should try to isolate and repair the problem as quickly as possible. When two or more ground faults accumulate on a single zone, they are very difficult to isolate, and sometimes the only remedy is to pull a new fire alarm cable. Faults to ground can be caused by a very slight pinching or creasing of the alarm cable, anomalies that would not affect a telephone or power circuit, so good workmanship in the original installation and any subsequent rework is key to trouble-free operation.

The third part of the system is made up of the indicating appliances. They consist of horns, chimes, strobe lights, and other equipment designed to alert occupants when the system has entered the alarm state. Because they draw more current than the initiating devices, heavier conductors may be needed depending on the number of indicating appliances on the zone. This zone is also supervised by the control panel, and there is also an end-of-line resistor, although its value is different, so care must be taken when making a replacement.

(Various manufacturers also employ different end-of-line resistor values. In Europe, end-of-line capacitors are used. In New Zealand, a manufacturer has come out with a control panel that can be programmed to work with either device. In the United States, the end-of-line resistor is located inside the junction box for the final initiating device or indicating appliance. In Canada, it is required to be installed in a separate enclosure.)

The fire alarm system is connected to the sprinkler system. If one of the heads ruptures as a result of heat below it, or if there is a break in a pipe, the water flow is detected by the control panel, and the system enters the alarm state. The alphanumeric display will report which sprinkler zone is affected. Every sprinkler head becomes, in effect, an initiating device.

The elevators are also integrated with the fire alarm system. Generally, the sprinkler system will immediately extinguish the fire. However, owing to an abundance of fuel supply and intensity of heat, it is possible that the fire would persist. Moreover, water from the sprinklers could short out the call button, a normally open switch, bringing the car to the affected floor. This would be the worst thing that could happen. As a protective measure, when there is a fire alarm in the vicinity of an elevator, the elevator system enters Phase 1 emergency recall. This compels the cars to return nonstop to a safe location, usually the ground floor. If the alarm condition originated from an indicating device at that location, the car goes to a previously designated alternate floor.

Phase 2 firefighters' operation takes over after the responders have arrived, and they have inserted a key into the lock switch that is identified by a red firefighter's hat. At that point, Phase 2 overrides Phase 1 so that the responders can take control of the elevator and use it in their effort to battle the flames.

Firefighters can manually control the motion of the car and operation of the doors. They know not to go to an affected floor and open the door, exposing themselves to intense heat and flames.

TIP ***Electricians who work in a facility where there is an extensive fire alarm system should make copies of the installation and operator's manuals and go through them carefully so that when the system goes into the trouble state, they will be prepared to cope.***

QUIZ

These questions are intended to test your comprehension of Chapter 6. The passing score is 70 percent, but try to answer them all correctly. The quiz, like most electricians' tests, is open book, so feel free to refer to the text. Answers appear in Answers to Quizzes and *NEC* Practice Exam.

1. **Flexible cords and cables may be used for**
   A. pendants.
   B. wiring of luminaires.
   C. elevator cables.
   D. Any of the above

2. **Flexible cords and cables may be used as a substitute for fixed wiring of a structure.**
   A. True
   B. False

3. **Three-way switches**
   A. are always used in groups of three per circuit.
   B. have one common terminal.
   C. are dangerous and should not be used except where absolutely necessary.
   D. are designed to protect the user from shock.

4. **To power down an entrance panel,**
   A. disconnect the conductors at the weatherhead.
   B. pull the meter.
   C. disconnect the ground rod.
   D. Any of the above

5. **When changing out a residential entrance panel,**
   A. pull the branch-circuit cables out of the connectors.
   B. be sure that the meter is in place.
   C. remove the connectors from the panel.
   D. wire temporary lighting from the meter socket lugs.

6. **Healthcare facility wiring involves**
   A. redundant grounding for patient-care areas.
   B. an alternate power source.
   C. transfer switches.
   D. All of the above

7. **Exposed conductive surfaces in a healthcare facility should be painted to eliminate shock hazard.**
   A. True
   B. False

8. **Class 1, 2, and 3 wiring**
   A. is very hazardous and always requires rigid metal conduit.
   B. always carries low voltage.
   C. is a useful alternative to standard wiring methods covered in *NEC* Chapters 1 through 3.
   D. is used for all communication circuits.

9. **Abandoned Class 2 wiring**
   A. is a fire hazard.
   B. must be removed.
   C. is capable of generating a large amount of toxic smoke.
   D. All of the above

10. **Fire alarm systems must be supervised at all times by trained professionals.**
    A. True
    B. False

chapter 7

National Electrical Code *Chapter 8: What's Different about Communications Systems?*

National Electrical Code (NEC) Chapter 8, "Communications Systems," is unique in the way that it relates to the rest of the Code. Unlike Chapters 5 through 7, which may in specific instances exclude themselves from some of the individual requirements of Chapters 1 through 4, Chapter 8 automatically excludes all provisions of those chapters unless they are specifically referenced. You could say that Chapter 8 is more autonomous or independent from the rest of the Code than any other chapter. That is fitting because its subject matter is also unique.

CHAPTER OBJECTIVES

In this chapter, you will

- Learn *NEC* provisions for communications systems.
- Discover the properties and work procedure for coaxial cable.
- Find out about characteristic impedance.
- Learn the fundamentals of satellite dish transmission and reception.
- See an overview of fiber-optic cable.

Article 800, "Communications Circuits," is an overview of the rest of the chapter. The primary topics are overhead clearances, underground conductors entering buildings, protective devices, grounding and bonding, cable hierarchy, and separation from other conductors.

As with all low-voltage cabling, there are the recurring sections concerning access to electrical equipment behind panels designed to allow access, mechanical execution of work, abandoned cables, and permitted cable substitutions.

Protective Devices

Part III describes protective devices. It states that a listed primary protector is to be provided on each circuit run partly or entirely in aerial wire or cable not confined within a block. Whenever there is exposure to lightning or the possibility of contact with outside power lines, there should be a primary protector. The Code provides grounding methods for the protector and for metallic members of the cable sheath.

Grounding specifications are in Part IV, which should be consulted if you are doing interbuilding work. Remember that the *NEC* does not have jurisdiction over utility work. However, if the individual premises includes two or more buildings and you are installing communications cabling between them downstream of the utility connection, *NEC* reenters the picture, and requirements should be observed.

The bonding conductor or grounding conductor is to be listed. It may be insulated, covered, or bare. It is to be copper or other corrosion-resistant conductive material, stranded or solid. It must not be smaller than 14 American Wire Gauge (AWG) and have a current-carrying capacity that is not less than the grounded metallic member and protected conductor of the communications cable. It is not required to exceed 6 AWG.

The primary protector bonding conductor or grounding-electrode conductor is to be as short as practicable. In one- or two-family dwellings, the primary-protector bonding conductor or grounding-electrode conductor is not to exceed 20 ft in length. It is to be run in as straight a line as possible. Any bend is similar to part of a turn in a wound coil, increasing the inductive reactance. This is significant for lighting, which with a very fast rise time resembles at that instant a high-frequency waveform.

Bonding conductors and grounding-electrode conductors are to be protected where exposed to physical damage. Where the bonding conductor or grounding-electrode conductor is installed in a metal raceway, both ends of the raceway are to be bonded to the contained grounding conductor or to the same terminal or electrode to which the bonding conductor or grounding-electrode conductor is connected. If the building served has an intersystem bonding termination, the bonding conductor is to be connected to it.

A bonding jumper not smaller than 6 AWG copper or equivalent is to be connected between the communication grounding-electrode and power grounding-electrode system at the building or structure served where separate electrodes are used. Bonding together of all separate electrodes limits potential differences between them and between their associated wiring systems.

Antennas and Satellite Dish Equipment

Article 810, "Radio and Television Equipment," covers antenna systems for radio and television receiving equipment, amateur and citizens-band radio transmitting and receiving equipment, and certain features of transmitter safety. The article also includes antennas, such as wire-strung, multielement, vertical rod, and dish types. It also covers the wiring and cabling that connects them to equipment.

Satellite dish receivers for television and Internet access are covered in Article 810. A satellite dish is actually a type of antenna. Dish installers are not required to be licensed electricians in most jurisdictions, but the overall installation must comply with the *NEC*, particularly in regard to grounding. Over half the installations are deficient, grounding and removal of abandoned cable being key issues.

Most electricians are not motivated to become satellite dish installers, but it is definitely beneficial to know how these systems work. From the transponders within the geosynchronous satellites in the sky above to the modem inside the building, dish systems bring together a variety of technologies, especially in the ways they cope with the high frequencies involved in getting the signal out of the studio, up into outer space, and then in encoded form down to the user's receiver.

Microwave reception requires line-of-sight access. It doesn't work to bounce the signal off the ionosphere as in the much lower-frequency AM radio transmission. Because most receiving locations would not have line-of-sight access, the most workable solution has been to broadcast the signal up to a satellite and then rebroadcast it back down to Earth. So that elaborate tracking mechanisms are not needed, geosynchronous satellites are used. They remain in a fixed location in the sky because their period of revolution about Earth is exactly equal to

Earth's period of rotation. A number of satellites are required to provide worldwide coverage. Each satellite has a unique footprint. For specific locations in areas within each of these zones, the apparent position of the satellite within the sky varies, so it is necessary to get aiming coordinates for the dish based on your zip code.

What makes it all work is the parabolic shape of the dish. A reflector whose shape is based on the parabolic curve has two unique properties. It will collect energy, light, or sound radiating from a remote source and deliver it to a single focal point. Likewise, it is able to collect energy from a focal point and radiate it in a narrow beam toward some remote location.

Still Struggling

A parabola is one of the conic sections generated when a plane intersects a cone. Its properties were known to ancient mathematicians. A parabolic reflector is also used for shaping the light from a parabolic bulb.

When a satellite dish receives a television or Internet signal from a satellite, it reflects it to the end of the feedhorn held by a mounting bracket so that it is at the focal point of the parabolic reflector (Figure 7-1). The signal from the satellite, as reflected to the feedhorn, is microwave, meaning that the wavelength is very short and the frequency very high—much higher than we encounter in conventional radio and television electronics. The frequency is so high that the signal cannot be transmitted via cable. Even the slight amount of capacitance in a short length of coaxial cable would be sufficient to shunt out the signal so that it could not survive the short journey to the edge of the reflector. It is possible for the signal to be conveyed in a different manner because the feedhorn is actually a structure that has the intriguing name *waveguide*. The waveguide is a hollow rectangular tube with a highly reflective inner surface, no moving parts, and dimensions that depend on the frequency of the signal to be conveyed. Because there is almost total reflection at the inner walls, the microwave signal is able to bounce off them and, following a zigzag path, travel the length of the feedhorn with almost no loss.

At the far end of the feedhorn is the low-noise block (LNB). This is the most elaborate and expensive part of the dish assembly. The signal is introduced into the LNB, which by means of a local oscillator creates a frequency

FIGURE 7-1 • Television satellite dish with feedhorn.

that will combine with the microwave signal to produce a resulting beat signal at a much more manageable frequency so that it can travel via coaxial cable 100 ft or whatever is required to reach the modem (modulator-demodulator).

The feedhorn is necessary because the LNB cannot be located at the focal point of the parabolic reflector. It would cast too big a shadow, blocking the microwave radiation.

PROBLEM 7-1

Why is satellite dish transmission necessary?

SOLUTION

At the frequencies necessary for bandwidth capable of conveying high-quality video, all transmission has to be line-of-site. Given the curvature of the Earth, this is made possible by transmitting to and rebroadcasting from a high-altitude geostationary satellite.

Many people assume that a television satellite dish transmits as well as receives. This is not true. It receives only. Any perceived interactivity takes place inside the receiver (not the television receiver, but the dish receiver, which resembles a cable box and sits adjacent to the television).

An Internet access dish, on the other hand, receives and transmits. This is why there are two coaxial cables from the modem to the dish or one twin cable. For this reason, care must be taken not to work on an Internet access dish while it is powered up, except for aiming. You can suffer radiation burns when the system is transmitting. The best way to ensure that the system is powered down is to unplug the modem and make sure that no one will plug it in while you are working at the dish.

Also, no system, television or Internet, should be powered up while you are terminating or working on the coaxial connections. This is so because if you inadvertently short out the coaxial cable, as would happen when cutting it, solid-state components would be damaged as a result of high current through the output circuits.

The coaxial cable between the building and the dish, in addition to the signal, also conveys a dc voltage that is necessary to power the solid-state components within the LNB. This dc voltage originates in the modem, and if it is interrupted, the system will not work.

Coaxial Cable

Article 820, "Community Antenna Television and Radio Distribution Systems," is the single *NEC* article that is all about coaxial cable. It defines coaxial cable as a cylindrical assembly composed of a conductor centered inside a metallic tube or shell, separated by a dielectric material, and usually covered by an insulating jacket.

As with other low-voltage cables, coax is subject to mandates concerning removal of abandoned cable, access to electrical equipment behind panels designed to allow access, mechanical execution of work, and cable hierarchy (i.e., plenum, riser, general purpose, restricted use).

The inner conductor, insulating layer, outer conductive shield, and jacket all share the same axis, which accounts for its name. Oliver Heaviside, a British theoretician and electrical researcher, patented coaxial cable in 1880. Several years later, additional technical advances, including waveguide transmission, were made. It wasn't until the 1930s, however, that real progress occurred. Closed-circuit television coverage of the Berlin Olympics was conveyed to

Leipzig over coaxial cable. In Australia, a 330-mile underwater coaxial cable carried broadcast and telephone signals. AT&T constructed a coaxial line in 1941 between locations in Wisconsin and Minnesota. It could carry either one television channel or 480 telephone circuits.

An important and useful property of coax is its ability to convey simultaneously a direct-current (dc) or alternating-current (ac) power and a radiofrequency (RF) signal. Electrical power is conveyed in the normal mode—a hot voltage on the inner pin and neutral or return voltage at ground potential on the outer shield. At the same time, the RF signal is conveyed in the transverse magnetic mode by means of magnetic and electrical fields outside the actual conductors, unlike conventional electric current, which is confined within the wires. These unlike forms of energy are conveyed simultaneously by the coax without detriment to either.

Cables that are designed to carry high-frequency signals exhibit a parameter known as *characteristic impedance*. Lower-frequency transmission lines such as ac power and telephone cables also have characteristic-impedance issues if they are sufficiently long. In Internet access and satellite television dish systems, characteristic impedance is important for the coaxial line as well as for unshielded twisted-pair (UTP) cable where it runs from the modem to the computer(s) with any intervening Ethernet hubs.

The most frequently used coaxial cable has a characteristic impedance of 75 Ω. This is a puzzling concept because it implies that if you were to measure with your ohmmeter the resistance for some specified length of the cable with or without the far ends shunted, you would get a reading of 75 Ω. This is not the case, though. The 75-Ω or other figure is valid regardless of the length. Characteristic impedance cannot be determined by means of an ohmmeter. To measure it, you need to use a very expensive instrument known as a *time-domain reflectometer* (TDR). Alternatively, characteristic impedance can be calculated using an intricate formula involving the properties of the cable—conductor size, spacing, and the dielectric constant of the insulating material. But you will not have to worry about the TDR or the calculations. In the ordinary course of coax installation, the electrician accepts the word of the manufacturer and goes by the marking on the cable. Still, it is helpful to understand what characteristic impedance is all about. It becomes relevant when high-frequency signals are transmitted over a pair of wires. Two conductors exhibit parallel capacitive and series reactance (Figure 7-2).

The two parallel conductors are equivalent to the plates of a capacitor, making for capacitive reactance. Inductive reactance results because the current

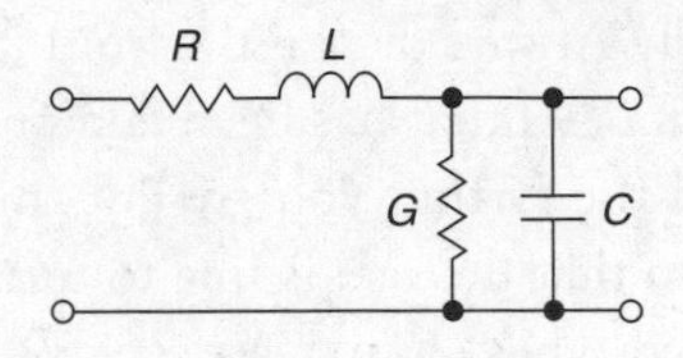

FIGURE 7-2 • Schematic of idealized transmission line.

flowing in the wires is fluctuating. The magnetic flux surrounding the two conductors is always moving. This work requires power, which has to come from somewhere, so the current flowing through the conductors is reduced, this effect appearing as inductive reactance.

If we imagine the two conductors to be infinite in length and divided, for the purpose of my demonstration, into an infinite number of uniform segments, we will see that the parallel capacitive reactance and series inductive reactance balance one another so that the impedance remains constant. This quantity is what constitutes the characteristic impedance of the cable. The length does not matter. Is the model flawed by the fact that in the real world the cable would not be infinite in length? No. Any cable that exhibits characteristic impedance will behave as if it were infinite in length if it is terminated in a load that has the same impedance. It is just about always necessary to carefully match impedances of media and load unless a mismatch is purposely introduced in order to diminish the signal for some reason. The ordinary procedure is to match impedances. If they are not matched, perhaps because of a kink or flaw in the cable, there will be harmful reflections that can cause data collisions, corruption, or other mischief.

Impedances also must match at the source and must be uniform all along the transmission line. This is why quality control in manufacturing and careful storage and handling at installation are necessary.

TIP ***When running coax, take care not to pinch or kink it. For high-frequency transmission, any damage to the cable will alter the characteristic impedance.***

Both inside and outside of equipment, where high frequency is involved, characteristic impedance plays an important role, and matching must be maintained. Even the traces on a printed circuit board exhibit characteristic impedance if the frequency is sufficiently high. In repairing equipment that operates at high frequencies, wiring should not be modified or rerouted so as to alter the characteristic impedance.

Coax is very user-friendly. Always dispense it from the original carton or reel. It should not be pulled off a reel that has been laid on end. This will make the cable increasingly twisted the farther you go. Put the reel on a metal rod or dowel held at either end so that the reel is free to turn.

The two commonly used types of coax are RG 59, used for short distances only, such as jumpers for consumer video components, and RG 6, which exhibits less loss at high frequencies and is good for longer distances. Diagonal cutters, a hacksaw, or tin snips can be used to cut coax. Shape the end so that it is not out of round. Make sure that the other end is not connected to equipment that is powered up.

The end can be prepared using a wire stripper and knife, but a coax stripper is faster and does a more precise job. The coax stripper makes both cuts simultaneously without cutting into the center conductor. Turn the cutter until both pieces are cut to the required depth; then remove scrap cutoffs. Fold the attached shielding back over the jacket to prepare the cable for insertion into the connector. Next push the cable all the way into the connector, making sure that the inner pin is centered and flush with the end of the connector. Then insert the prepared end of the coax into the crimping tool. Squeeze the handle of the crimping tool, and you are done. The whole process takes very few seconds and produces a better termination than the old die-type crimper.

You can connect the coaxial cable to equipment using the free-turning threaded part of the connector. Don't use a wrench, just put it on hand-tight. Take care not to turn the fitting that is built into the equipment. This would damage electronic components inside the housing.

Article 830, "Network-Powered Broadband Communications Systems," states in Section 830.1, "Scope," that this article covers communications systems that provide any combination of voice, audio, video, data, and interactive services through a network interface unit.

Article 830 systems are similar to Article 820 systems, but with a greater array of services provided, more elaborate electronics are necessary. Therefore, a higher voltage level is permitted, 150 V as opposed to the 60-V maximum permitted for Article 820 networks.

As in other communications systems' articles, there are provisions that relate to abandoned cable, access to electrical equipment behind panels designed to allow access, mechanical execution of work, and cable hierarchy.

Article 840, "Premises-Powered Broadband Communication Systems," also has similar requirements. This article is applicable when fiber-optic cable is used to bring subscriber services to the premises. Power for the optical network terminal, which is considered to be network interface equipment, is from within the customer's premises.

Optical Fiber

Electricians are divided when it comes to optical fiber. Some have embraced this newer technology, and they see it as a way to secure more work in an uncertain economy. Others feel that a line has to be drawn somewhere and that optical fiber should remain outside the scope of the profession.

Copper wire, notably unshielded twisted pair (UTP), is adequate for today's needs, but as time passes, our expectations with regard to network speed and capacity are increasing. Many industrial and commercial facilities contain at least some optical fiber, and it is likely that the proportion will increase. Another factor powering the move to fiber is the fact that there is the potential for its price to fall, whereas copper prices may continue to rise. Regardless of your current feelings on the subject, it is undoubtedly beneficial to understand how this interesting technology works, so I will provide at least an overview.

Optical fiber has these advantages over copper media:

- Isolation between input and output is total.
- For noncomposite optical fiber, there is no danger of electric shock or fire initiation.
- Over long distances, there is greater bandwidth.
- Repeaters for boosting a weakened signal are fewer and farther apart.
- Less labor is needed to run optical fiber because it has a smaller diameter and is lighter.
- For optical fiber, there is no electromagnetic interference. It is more suited to noisy industrial settings with motors and fluorescent ballasts, and there is never crosstalk.

PROBLEM 7-2

Why is optical fiber able to convey greater bandwidth than copper wire?

SOLUTION

Light has a higher frequency than radiofrequency electric current, so it can be modulated or digitized at a higher resolution.

In contrast to the flow of electrons that constitutes electric current in copper wire, optical fiber carries pulses of light. The fiber is yet another variety of waveguide. The dimensions of any waveguide depend on the frequency, and hence the wavelength, of the signal to be conveyed. Light at very high frequencies requires a narrow waveguide, so the diameter of an optical fiber is of the same order of magnitude as that of a human hair.

Unlike other waveguides, an optical fiber is not hollow. The inner core, however, has a higher index of refraction than the outer cladding. Consequently, the light bounces from side to side of that interface, remaining confined to the inner core with very little loss.

Still Struggling

Naturally occurring thermal layers in the ocean make huge waveguides that are capable of carrying the low-frequency voices of whales over great distances. The sound waves bounce off the boundaries between layers and travel a zigzag path with little loss.

There are two different types of optical fiber: single mode and multimode. Multimode fiber has a comparatively large core diameter, up to 100 μm, significantly larger than the wavelength of the light that it transmits. Being larger, it is easier to terminate and overall more user-friendly. But there is a tradeoff. Because it carries light of multiple wavelengths simultaneously, and because these signals do not all propagate at the same rate, distortion is inevitable in proportion to the length of the cable run. For this reason, multimode fiber is only used for short indoor runs, whereas the more challenging single-mode fiber is used for utility-scale work and longer interbuilding runs. If high-speed performance is to be maintained, multimode fiber has a maximum length limit of about 1,000 ft. For single-mode fiber, the signal source is laser; for multimode fiber, it is usually the simpler light-emitting diode (LED). Optical fiber can be identified by the color of the jacket. Multimode is orange and single-mode is yellow.

If you as an electrician have decided to go into optical fiber, you will most likely confine your operations to multimode. Outside work using single-mode optical fiber is performed by utility workers or high-tech outside contractors.

To do this work, they need, at a minimum, a temperature-controlled van in which to make fusion splices; an aerial bucket truck, a backhoe, or an excavator; and a flatbed for hauling it. Additionally, much test equipment is needed, including the $5,000 optical time-domain reflectometer (OTDR) with launch and receive cables.

If you narrow your scope to indoor work using multimode fiber, you will not need to do splices. If a run of optical fiber is damaged or does not perform well, it is replaced in its entirety.

As with all wiring, optical fiber is a complex technology, and a certain amount of background learning and on-the-job experience are needed to enter the field. Internet and print resources abound. An excellent website is www.thefoa.org, maintained by the nonprofit Fiber Optic Association. The site contains a wealth of free technical information leading to certification.

QUIZ

These questions are intended to test your comprehension of Chapter 7. The passing score is 70 percent, but try to answer them all correctly. The quiz, like most electricians' tests, is open-book, so feel free to refer to the text. Answers appear in Answers to Quizzes and *NEC* Practice Exam.

1. *NEC* Chapter 8

A. automatically excludes all provisions of Chapters 1 through 4 unless they are specifically referenced.
B. is subject to provisions in other Code chapters.
C. is all about Class 2 circuits.
D. is subject to the overcurrent protection rules in Article 430.

2. Communications circuit conductors are subject to the cable hierarchy rules.

A. True
B. False

3. In one- and two-family homes, the length of a primary-protector bonding conductor is to be not over

A. 8 ft.
B. 10 ft.
C. 16 ft.
D. 20 ft.

4. Microwave reception requires line-of-sight access.

A. True
B. False

5. The feedhorn

A. contains the LNB.
B. is a waveguide.
C. in a television system may be a source of radiation burns.
D. is prone to malfunction due to its many moving parts.

6. The coaxial cable between dish and building

A. must be no longer than 30 ft.
B. conveys a dc voltage.
C. is very fragile.
D. can be replaced by telephone wire.

7. The most frequently used coaxial cable has a characteristic impedance of

A. 50 Ω.
B. 75 Ω.
C. 100 Ω.
D. 150 Ω.

8. Optical fiber is a type of waveguide.

A. True
B. False

9. Optical fiber has which of these advantages over copper media?

A. Isolation between input and output is total.
B. Over long distances, there is greater bandwidth.
C. There is no electromagnetic interference.
D. All of the above.

10. Multimode optical fiber is more difficult to work with than single-mode optical fiber.

A. True
B. False

chapter 8

National Electrical Code *Chapter 9:* *Working with Tables*

As electricians become more familiar with the *National Electrical Code* (*NEC*), they spend less time reading text and focus more on the tables. The text lays down the basic definitions and principles, and the tables supply the exact numbers. With practice comes familiarity, and the most-used areas become part of our knowledge base. But none of us can remember with certainty all the figures that are in the tables, especially in as much as they are subject to change in successive *NEC* revisions that appear every three years.

CHAPTER OBJECTIVES

In this chapter, you will

- See how tables help electricians to design their installations.
- Learn about sizing conduit and tubing runs.
- Find out how to determine ampacities and choose conductor sizes.
- Take a look at the Annexes.

In this chapter, I will go over some of the more important tables. They appear in Chapter 9, "Tables," in the Annexes and spread throughout the Code.

Conduit Size

Many of the tables have to do with sizing conduit and tubing fill. By this is meant the maximum number of conductors of various sizes that can be installed in a run of raceway. If you attempt to overfill the pipe, the wires may not go at all, or if they do, so much force will be required that the insulation may be damaged, or the conductors may overheat when loaded to close to capacity because more heat is generated than can be dissipated. If conduit and tubing fill limits are observed, there will not be a problem, but to determine the correct amount, a series of calculations is required, making reference to the applicable tables.

These calculations should be performed as part of the planning stage in any raceway installation. Pipe jobs, including materials and labor, constitute a significant investment, especially in the larger sizes or if the conduit is rigid metal. It is a waste of resources if this pipe is installed only to find that it is undersized for the number, size, and usage of the conductors to be pulled.

TIP ***Before starting a large raceway installation, make a conduit and tubing schedule showing number of runs and sizes.***

Fill requirements can be met by using a sufficiently large raceway or by employing more than one run. If the latter, a circuit consisting of hot wire(s), neutral, and equipment ground should not be separated but run in the same raceway, both to expedite future circuit tracing and to prevent inductive heating.

Conduit and tubing fill figures as given in the tables are valid only if certain principles are observed. For one thing, a raceway system is to be installed as a complete system and terminated at both ends before the conductors are installed. This is as opposed to sliding the individual raceway segments onto the conductors. A pull rope can be put in place that way, but the more professional method is first to complete the raceway run including backfill if it is an underground line. Then put your *mouse* to work. This is a reusable rubber cylinder sized to fit the raceway, also called a *piston*. It has a metal hook, and light string can be attached. Then, at the far end or at each intervening pull box, use a shop vacuum to pull the mouse, trailing the string, through the raceway. The string is then used to pull the heavier rope that is needed to pull the conductors. Instead of attaching string to the mouse, you can use a light cloth tape that is made for the purpose. It is marked in feet and is a great aid in determining the amount of wire that will be needed.

There is no maximum length to a raceway run, but it must have no more than the equivalent of four 90-degree bends, including sweeps and offsets, between pull boxes. If it looks like it might be a difficult pull, put in more pull boxes.

PROBLEM 8-1

Where can raceway specifications and installation requirements be found?

SOLUTION

Toward the end of *NEC* Chapter 3 is a series of articles on individual raceway types. Of particular interest are sections on "Uses Permitted" and "Uses Not Permitted."

Water-pipe fittings are compatible with conduit threads [or for polyvinyl chloride (PVC), they fit the conduit], but they should never be used, nor should water pipe be substituted. The inside of galvanized steel water pipe is not as smooth as RMC and is sure to damage the conductors in a heavy pull. Close conduit fittings such as 90- and 45-degree bends have removable access panels to aid in pulling wire.

All bends in a conduit or tubing run are to be factory-made sweeps or bends made in the field using a hand or power conduit bender. These tools make gradual, uniform bends of the required minimum radius, and the inside diameter of the pipe is not materially reduced.

The raceway run should be kept away from sources of heat such as steam or hot water pipes or heat ducts.

Conduit and tubing should not be painted or covered with any material that could interfere with heat dissipation. Power and low-voltage cabling should not be hung from the raceway because this could add weight and impede heat dissipation.

Various types of raceways are covered in individual articles in *NEC* Chapter 3. The requirements for a contemplated installation, especially "Uses Permitted" and "Uses Not Permitted," should be reviewed prior to planning an installation.

As far as sizing an installation is concerned, a number of tables are applicable. First, ascertain the number and sizes of the conductors that are to be installed in the raceway.

Referring to Table 310.15(B)(3)(a), "Adjustment Factors for More than Three Current-Carrying Conductors in a Raceway or Cable," decide how many of these there will be. Any hot wires that can be loaded concurrently have to be counted as current-carrying. Additionally, the neutrals that may carry current other than the difference between two similarly loaded ungrounded conductors must be counted as current-carrying. Equipment-grounding conductors do not have to be counted as current-carrying. If there is any doubt about the status of a neutral, consider it current-carrying.

You will see in the table that if there are more than three current-carrying conductors in a raceway, various percentages must be applied to the allowable ampacities found in Tables 310.15(B)(16) through 310.15(B)(19). These percentages will adjust the allowable ampacities in a downward direction. The amounts differ depending on the total number of conductors (not just current-carrying) in the raceway. If there are four to six such conductors, the adjustment factor is 80 percent. At the far end of the scale, if there are 41 or more such conductors in the raceway, the adjustment factor is 35 percent.

The conductors that are counted are power and light conductors only. Remote-control, signaling, and power-limited circuit conductors do not have to be counted as current-carrying, and the ampacity adjustments apply only if they carry continuous loads in excess of 10 percent of their ampacities.

For ambient temperatures, there are additional correction factors. Table 310.15(B)(2)(a) provides these factors based on 86°F. In other words, if the ambient temperature is less that 86°F, the correction factor is greater than 1.00, the ampacity of the wire may be increased. If the ambient temperature is greater than 86°F, the ampacity of the wire is decreased, in both cases referring to Tables 310.15(B)(16) through 310.15(B)(19).

Table 310.15(B)(2)(b) provides the same information based, however, on an ambient temperature of 104°F.

It is to be emphasized that for common indoor installations, the ambient temperatures have less to do with the outdoor climate than with heat sources in the vicinity of the circuit in question. Care must be taken when passing through a boiler room, for example, to take note of any elevated temperatures than may be encountered.

TIP ***Adjustment factors and correction factors must be applied prior to sizing conductors and raceways. The order in which they are applied does not matter.***

There is another table that is relevant in some cases. It has been found that directly over a roof, circular raceways tend to heat up to a great extent when they are in direct sunlight. In effect, the ambient temperature becomes higher. Therefore, in Table 310.15(B)(3)(c), temperature adders are given for different heights above the roof. These amounts are added to the outdoor ambient temperature. A source for determining outdoor ambient temperatures in various locations is the *ASHRAE Handbook: Fundamentals*.

All this may seem rather complex, but if each subtopic is approached in an orderly fashion, it will all come together. Working electricians perform these calculations many times, and after a while, they become quite familiar. Keep in mind the fact that we are working toward sizing the raceway system, and to do so, we have to generate a list of all conductors, including sizes, that are going to be installed in it.

Having completed the preceding determinations, you are prepared to make a list of conductors that will go in each raceway. For this task, refer to Tables 310.15(B)(16) through 310.15(B)(19). Tables for higher voltages appear later in this article, but these are the most used tables for premises wiring. Each table has a title that explains its application, that is, whether the conductors are in raceways or free air, supported on a messenger, and so on. By far the most used is the first table. Most licensing exam questions and real-world installations are based on this table, which covers allowable ampacities of insulated conductors rated up to and including 2,000 V, 60 through 90°C, not more than three current-carrying conductors in a raceway, cable, or earth (directly buried) based on an ambient temperature of 86°F.

The *NEC*, where referring to temperature, provides both Fahrenheit and Celsius amounts. Because the United States has not yet gone metric, in this book I have been quoting the Fahrenheit amounts. That is what you will see in most licensing exam questions. However, it is an easy matter to work in either scale. For conductor temperature ratings, it is customary to use the Celsius scale. For example, the much-used THHN is rated 90°C, whereas UF is rated 60°C.

Notice that this table has the range of conductor sizes in the left-hand column beginning with American Wire Gauge (AWG) and shifting to kcmil for sizes over 4/0. The sizes are repeated in the right-hand column for ease in reading.

Across the top are temperature ratings of the various conductors, and under each figure is a list of the specific wire types.

The left side of the table is for copper conductors, and the right side is for aluminum or copper-clad conductors. Most licensing exam questions refer to copper, and this is the most frequent installation, so usually you will be looking at the left side of the table.

After correction factors have been applied, pick out the corrected ampacity, and note the conductor size. Now you can make a list of the conductors (including sizes) that will go in the raceway, in preparation for sizing the conduit or tubing.

The conductors intended to go in the raceway are very often all the same size, and in this case, the procedure for sizing the raceway is quite simple. Just turn to Annex C at the back of the Code. Tables C.1 through C.12 give maximum numbers of conductors that can be installed in raceways only when all conductors are the same size. These tables are completely user-friendly. Each table applies to a different type of raceway. For example, consult Table C.1 if you are using EMT. In the left-hand column are conductor types, with sizes in the second column. Across the top are raceway sizes, provided in both metric designators and trade sizes. The body of the table gives the maximum number of conductors that can be installed. For example, if you are working with 12 AWG THHN and you have 14 conductors, you will see that ¾-in. EMT is required.

As you can see, as long as all conductors are the same size, as is often the case, the conduit or tubing fill is easy to determine. If the proposed installation is a mix of sizes, the calculation is more intricate. However, it is sometimes possible to simplify the procedure. For example, if you have twelve 10 AWG THHN conductors and four 12 AWG THHN conductors, you could look up the raceway size for sixteen 10 AWG THHN conductors, and you would be sure that you were not undersizing the raceway. On a borderline installation, this procedure could trigger a larger raceway than required, so you would have to decide if that would be objectionable from a cost standpoint. In a licensing exam, this procedure could conceivably give a wrong answer, but most of the time you could tell by looking at the multiple-choice alternatives.

If there is a mix of conductor sizes, which is more common in licensing exams than in actual installations, a different procedure for sizing the raceway becomes necessary. You have to find the cross-sectional areas of all conductors and the cross-sectional area of the raceway. Then divide the former into the latter to see if the conductors will go. In Chapter 9, Table 5 gives dimensions of insulated conductors, including their areas in square inches and square millimeters. Add to the list of conductors with sizes their cross-sectional areas. Make subtotals for the number of each size and type and a grand total for all conductors to be installed in the raceway.

Then consult Table 4 in Chapter 9. This table gives the total internal cross-sectional areas for various types of raceway, again beginning with EMT. In addition, 60, 53, 31, and 40 percent figures are given in separate columns in order to account for varying numbers of conductors in the raceway and for raceways segments that are not over 24 in. long. Because the percentages are built right into the table, it is not necessary to look them up, nor is it necessary to perform the calculation. Divide the conductor cross-sectional areas into the raceway cross-sectional areas on a trial basis until you find the required raceway size. The areas here again are given in square millimeters and square inches. You can use either, but choose the same for both conductors and raceway.

TIP ***The tables for finding cross-sectional areas of conductors and raceways are located together in NEC Chapter 9.***

This is how you size raceways for any installation. As mentioned earlier, when the conductors are multiple sizes, it is a little more difficult, but with a handheld calculator to do the number crunching, it is not overwhelming.

Chapter 9 contains some other tables. Table 1, "Percent of Cross Section of Conduit and Tubing for Conductors," gives three categories that are valid for all conductor types and for all raceways. If there is just one conductor, 53 percent fill is permitted. If there are two conductors, 31 percent fill is permitted. If there are more than two conductors, 40 percent fill is permitted. These are all maximum amounts. Notice that for two conductors the fill permitted is less than for one conductor or for over two conductors. The reason for this puzzling situation is that when two conductors are being installed in a raceway, their diameters are added, whereas if more than two conductors are being installed, they tend to arrange themselves in such a way as to use the available space more efficiently.

Still Struggling

Conduit fill specifications vary for different types of raceways in part because the inside diameters are not all the same.

Table 1 does not include the 60 percent category that we saw earlier pertaining to short lengths (24 in. or less) of conduit or tubing. Table 8, "Conductor Properties," gives dc resistance at 176°F for various size conductors. This information may be used to calculate voltage drop (using Ohm's law) and also to find the distance in an underground line to a short circuit, saving a lot of digging. Table 9 gives ac resistance for three-phase cables at 60 Hz.

Tables 11(A) and 11(B) give power limitations for Class 2 and Class 3 circuits. Tables 12(A) and 12(B) do the same for power-limited fire alarm systems. This information is used by listing organizations when they investigate power-limited sources, and it is not ordinarily needed by working electricians, who will instead consult the marking on the transformer or power source to see if it qualifies, for example, as Class 2.

The Annexes are a series of text sections with tables that stand outside the main body of the Code. Rather than containing mandates, they are informative only. Some of them you will rarely consult, whereas others are very useful.

Annex A provides a list of product safety standards used for product listing where the listing is Code required. Annex B provides application information for ampacities calculated under engineering supervision and is not part of the electrician's normal course of study.

Annex D, "Examples," is an excellent aid in studying the *NEC*. These examples illustrate methods for calculating ampacities in various occupancies. The material is intended to supplement Chapter 2 and other portions of the Code that are heavy on calculations. If you have a licensing exam in your future or a specific question on some application in the field, Annex D will be quite helpful.

Annex E, "Types of Construction," is interesting if you are attempting to identify a building type in order to see if Type NM (Romex) cable is permitted. Many of the better electricians, however, use metal raceways for all their indoor commercial and industrial work.

PROBLEM 8-2

Why is Annex E important?

SOLUTION

It is used to determine where Type NM (Romex) cable is permitted.

Annex F concerns critical-operations power systems (COPS). It supplements Article 708, which is part of Chapter 7, "Special Conditions." The concern in Annex F is to ascertain the availability based on number of hours of downtime in a year. The rationale is that because COPS may support facilities with objectives that are vital to public safety, any system downtime is costly in terms of economic losses, loss of security, or loss of mission.

Annex G, "Supervisor Control and Data Acquisition (SCADA)," concerns the reliability of these interesting systems. SCADA is a system for monitoring from a remote location an unmanned facility such as an electrical substation. Communication may be via dedicated telephone line(s) or the Internet. It is provided that where the SCADA system is managing a COPS facility, no single-point failure is to be able to disable the SCADA system. Power supply, security against hazards, and maintenance and testing are addressed in Annex G.

TIP ***SCADA is used for monitoring individual turbines in a wind farm.***

Annex H, "Administration and Enforcement," provides a model set of rules that is offered up for jurisdictions to adopt in whole or in part if they wish. Provisions include fees, inspections and approvals, inspector's qualifications, and so on.

Annex I, "Recommended Tightening Torque Tables," gives figures for tightening electrical connections of various types and sizes. We all know that electrical connections benefit from being sufficiently tight, and if they are left loose, there will be a palpable fire hazard down the road. But there is also such a thing as overtightening, and if this is done, the threads can be damaged so that the connection in time will become loose. Use of torque wrenches and torque screwdrivers in conjunction with these tables is the way to ensure top-quality connections, and it is helpful in getting a feel for what is required

QUIZ

These questions are intended to test your comprehension of Chapter 8. The passing score is 70 percent, but try to answer them all correctly. The quiz, like most electricians' tests, is open-book, so feel free to refer to the text. Answers appear in Answers to Quizzes and *NEC* Practice Exam.

1. **To size conduit or tubing,**
   A. there are alternate methods depending on whether the conductors are all the same size.
   B. Annexes, tables, and the main body of the Code are all relevant.
   C. adjustment and correction factors must be considered.
   D. All of the above

2. **A conduit run must be completed before the conductors are installed.**
   A. True
   B. False

3. **The equivalent of how many 90-degree bends can be in a conduit run without an intervening pull box?**
   A. Two
   B. Three
   C. Four
   D. Five

4. **Water pipe may be used as a substitute for rigid metal conduit.**
   A. True
   B. False

5. **If there are four to six conductors in a raceway and more than three are current-carrying conductors, the ampacity must be adjusted by**
   A. 40 percent.
   B. 50 percent.
   C. 60 percent.
   D. 80 percent.

6. **If all the conductors are the same size, conduit fill may be found by consulting**
   A. Annex A.
   B. Annex B.
   C. Annex C.
   D. *NEC* Chapter 3.

7. **If all conductors are not the same size, conduit fill may be found by consulting**
   A. Annex A.
   B. Annex B.
   C. Annex C.
   D. *NEC* Chapter 9.

8. Conductor properties are found in

A. Table 7.
B. Table 8.
C. Table 9.
D. *NEC* Chapter 2.

9. Information on COPS is found in

A. Annex A.
B. Annex B.
C. Annex C.
D. Annex F.

10. Information on SCADA is found in

A. Annex B.
B. Annex C.
C. Annex D.
D. Annex G.

chapter 9

Tools and the Job Site: Efficient Work Procedures

As we have emphasized, an essential item in the electrician's toolbox is the *National Electrical Code* (*NEC*)—both the physical book and the body of knowledge. Other tools are important as well, and we'll survey them in this chapter. Then we'll consider job-site procedures that promote efficiency and productivity.

CHAPTER OBJECTIVES

In this chapter, you will

- Survey electricians' measuring tools.
- Learn how to check diodes and capacitors with a multimeter.
- Find out about the clamp-on ammeter.
- Check out some loop impedance meters.
- See what is new in oscilloscopes.

As for tools, measuring instruments are needed for most troubleshooting and repair. We will start with the ubiquitous multimeter (Figure 9-1), so called because a single meter is able to measure at least three circuit parameters: volts, ohms, and milliamperes.

A cheap multimeter is adequate for many tasks. A really good model will cost over 10 times the amount noted in the figure caption, and in the long run, that is probably the one you want. With ruggedized case, extended ranges, and large, clear display, it will last a lifetime. Still more expensive models include extra features such as a temperature probe, diode and capacitor tester, network diagnostics, and graphic display. Depending on your budget, you can shop around and pick the model that is right for you.

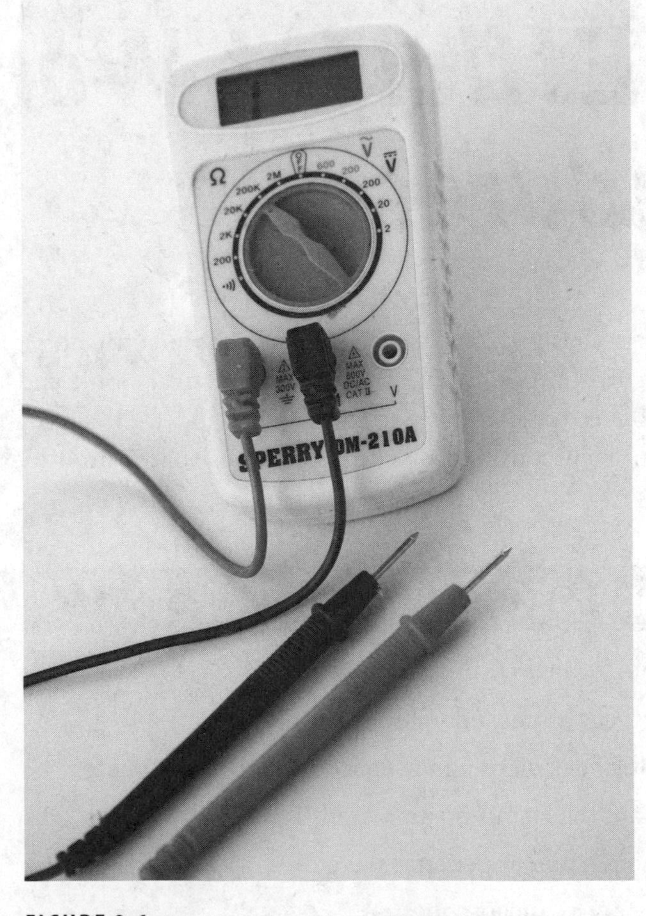

FIGURE 9-1 • Inexpensive multimeter bought at a big-box store for about $10.

Types of Multimeters

The multimeter is available in two versions: digital, with an alphanumeric display of great accuracy, and the traditional analog, with a needle that indicates the value relative to the correct scale. A mirror situated behind the needle ensures that you are not viewing the movement at an angle, which would distort the reading. Some electricians prefer an analog multimeter. A possible advantage is that it is more accurate at very low temperatures, but for most of us, the digital meter is the clear winner. Both models are widely available, and they come with good user manuals and guarantees. Most of the discussion that follows is applicable to the digital model. We will start with the volts function.

A modern digital multimeter in the voltage-reading mode is a very high-impedance instrument. When you connect the probes across a source or load so as to take a reading, your meter comprises a very high resistance. It draws an infinitesimal amount of current. This is a desirable feature because it means that if the source has significant internal impedance, the act of measuring will not load the circuit, drop the voltage, and give a false reading. The source voltage is not what powers the display (or moves the needle in an analog meter). That power comes from an internal battery. The source or circuit voltage that is being measured becomes the input for a field-effect transistor (FET) at the meter's front end, and this solid-state device requires almost no current to produce a relatively large output.

This property is valuable when diagnosing electronic equipment, but one of your most common tasks with a multimeter in the volts function is to check wires and terminals for the presence or absence of voltage where the exact amount is not an issue. Often you want to verify that a circuit is not live before working on it. For this test, it is essential first to probe a known live source just to make sure that you are not deceived by a dead battery in your meter or an open probe lead.

PROBLEM 9-1

What are the primary multimeter functions?

SOLUTION

The primary functions are voltmeter, ohmmeter, and ammeter.

Phantom Voltage

On first using a high-impedance digital multimeter in the volts function, you may be puzzled by the phenomenon of *phantom voltage*. The readout will appear to drift about at a very low level. Don't worry about it. As soon as you connect to something live, the meter will lock onto the real energy, and the readout will stabilize. The explanation is that static voltage and low-level transient currents are everywhere, in the meter circuitry, in the plastic housing, and in the probes. It is this energy that you are reading, and it happens with a high-impedance instrument. A low-impedance meter will not see phantom voltage because it will load the circuit and keep the potential drained out.

The ohm function is very useful in troubleshooting and repair, as well as in checking newly installed wiring to make sure that you don't have an unintended short. In the ohm function, the meter's probes should not be connected to a live circuit. Some meters will be instantly damaged if connected to a live circuit, whereas others are protected by an internal fuse. Full-featured instruments are not affected as long as the fault is of low level and short duration. Such instruments make an audible alarm, and you should disconnect as quickly as possible.

Many ohmmeters have a very desirable feature: a soft audible signal indicates continuity. It sounds whenever the ohmmeter sees less than 20 to 30 Ω. This is useful when you have to use care in placing the probes, and it is difficult to watch the readout at the same time. Continuity readings are frequently needed in troubleshooting and verifying the integrity of new wiring. For equipment that is not energized, I like to take an ohm reading between a line terminal and the metal enclosure to make sure that there is no ground fault. This is a good practice after rewiring old light fixtures.

Low-ohm readings become increasingly difficult to read as the resistance approaches zero because there is inevitable impedance where the probes contact the work, and this makes for an unstable, fluctuating reading. If a power tool or appliance will not run or runs intermittently, insert the ohmmeter probes through the holes in the plug prongs, and wrap them with electrical tape to maintain good electrical contact. Then press the trigger or power switch, watching the ohmmeter readout and listening for the continuity tone. Most working power tools will read low ohms when the motor is not turning, provided that the switch is on. Wiggle the switch from side to side, and flex the cord along its entire length, especially where it enters the housing. If the readout or audible continuity signal fluctuates, you've found the problem. If not, you can disassemble the tool with the meter still hooked in place. You should get continuity straight through to the armature with the switch on. If there is no continuity, shunt the circuitry at various points, and you should quickly find the fault.

If you are getting continuity but the motor won't run, take a high-range ohm reading from either of the line conductors to the metal enclosure. You should get a high megohm reading. If not, the windings are grounded out or there is a short in the wiring.

Ohmmeter readings are helpful for other tasks as well. Resisters, capacitors, coils, transformers, switches, and diodes are some of the components that can be easily tested. Transistors also can be checked, but the test is not totally definitive.

Testing In and Out of Circuit

Devices often can be tested in circuit, but if there is a parallel loop of sufficiently low impedance, the reading will not be valid. If you think that this is possible, disconnect one of the leads (of a two-lead device). If you desolder a lead and/or resolder it after taking the measurement, be careful not to damage the device by applying too much heat. Use the smallest soldering iron that will get the job done. Heat sinks, made for the purpose, are effective in absorbing excess heat. Solid-state devices are particularly vulnerable to heat damage

Resisters are usually color-coded. Polarity doesn't matter. Find the appropriate ohm range, either by trial and error or by reading the color code, and determine whether the resister is within tolerance. Most faulty resisters are open, so it is evident when they have gone bad, and of course, that will take down the circuit and often the entire piece of equipment. A bad resister will often have a characteristic charred appearance.

Testing a Capacitor

Capacitors also can be checked with an ohmmeter. The internal battery places a direct-current (dc) voltage across the probes, for most meters about 3 V. This will either charge or discharge a capacitor depending on its current state and the polarity of the hookup. If you put your meter in a medium megohm range and touch the probes to the capacitor's leads, you will see the resistance rise or fall in a distinctive steady manner, slowing down as the capacitor charge approaches an end state. Technicians call this behavior *counting,* and it is most pronounced for a high-capacity electrolytic capacitor.

TIP ***An electrolytic capacitor that has been out of use for a while may go bad. Sometimes it can be restored by applying a steady dc voltage.***

If you have an inquiring mind, you probably like to take discarded electrical equipment apart to see how it works. Often you can follow the power flow, starting at the power cord where it enters the cabinet and ending at the output, which may be a speaker(s), a motor, a light, or some sort of actuator. Disassembly is instructive also in showing how to open plastic enclosures without damaging or marring them. Measure the electrical parameters of the components, first in circuit and then with all but one lead disconnected, to see if there is a difference. You can save components for purposes of experimentation and trial-and-error substitution, but generally where reliability is a consideration, only new replacement parts should be used.

A word of caution: some electrical equipment contain hazardous and, under certain conditions, lethal voltages long after the unit has been powered down and disconnected from the power source. Large power-supply capacitors store a potent charge, and throughout a piece of equipment there may be distributed capacitance, which can be treacherous. Some technicians short the terminals with a wire to eliminate the charge. This can expose you to shock as well as arc-flash hazard. Also, it can damage the component, for example, by puncturing the dielectric layer of a capacitor. A much better method is to shunt any terminals that could have a high-voltage potential by connecting a power resister of appropriate value across the terminals so as to do a sensible discharge.

Testing a Diode

A multimeter also can be used to check a diode. In ohm function, the meter applies a dc bias to the component, and depending on the hookup, you will read high resistance with the probes hooked one way and low resistance when reversed. If your meter has a diode check function, use that instead. Rather than providing a pseudoresistance reading, it measures voltage drop both ways and gives a go, no-go report.

TIP ***Most resistors, capacitors, and diodes are very inexpensive, so they should be replaced if their quality is questionable.***

An ohmmeter can be used to test a great number of components, and it constitutes a quick and reliable starting point for troubleshooting equipment and circuits. You can check light bulbs (not fluorescent—they require an ionizing voltage to fire up), heating elements, switches, motors, and so on. When wiring a meter socket, entrance panel, or branch circuit, you can check your wiring with an ohmmeter before applying power.

Do you need to diagnose a long underground line? Unhook both ends, and do resistance readings, first with the conductors at the far end not connected and then shunted. You can find out if you have a short or open condition. If there is a short, you can estimate the location by reading the resistance from both ends.

Another multimeter function is the milliammeter mode. This is rarely used. To measure current flow, you have to break open the circuit and insert the meter in series with the load and power source, the equipment being powered up. The entire current flows through the meter as opposed to the volt mode, where the meter is placed in parallel with the source or load, and a much smaller amount of current flows through the meter. In milliamp mode, it is easy to overload the meter.

Using the Clamp-On Ammeter

When there is significant current flow, such as at a service or big load (e.g., a hot water heater or motor), a much safer and more accurate way to measure current is by means of a clamp-on ammeter. This instrument consists of a pair of spring-loaded jaws and a digital or analog readout. You enclose within the jaws the conductor in question, with the meter set to the appropriate range, and read the result. The clamp-on ammeter is sensitive to the magnetic field that surrounds any conductor when current is flowing through it. This instrument gives a good, stable reading, and it makes no difference whether the conductor is centered within the jaws or passes through them at an angle rather than exactly perpendicular. But you have to realize that there must be only one conductor passing through the opening. If there are two conductors of opposing phase, such as the two legs of a 240-V single-phase circuit, the currents will be flowing in opposite directions, and they will cancel out, giving a reading of 0 A. The same is true if there is one hot wire and its associated neutral, as in a 120-V single-phase load. For this reason, it is not possible to read the current flow in a cable containing two or more conductors or in a cord. You can open up the equipment and find a single wire or go into a junction box or entrance panel. Another method is to make a splitter (Figure 9-2). Put a plug and connector on either end of a 2-ft flexible cord. Slit about 8 in. of the jacket near the middle of the cord without nicking the insulation of the wires. Pull out the conductors,

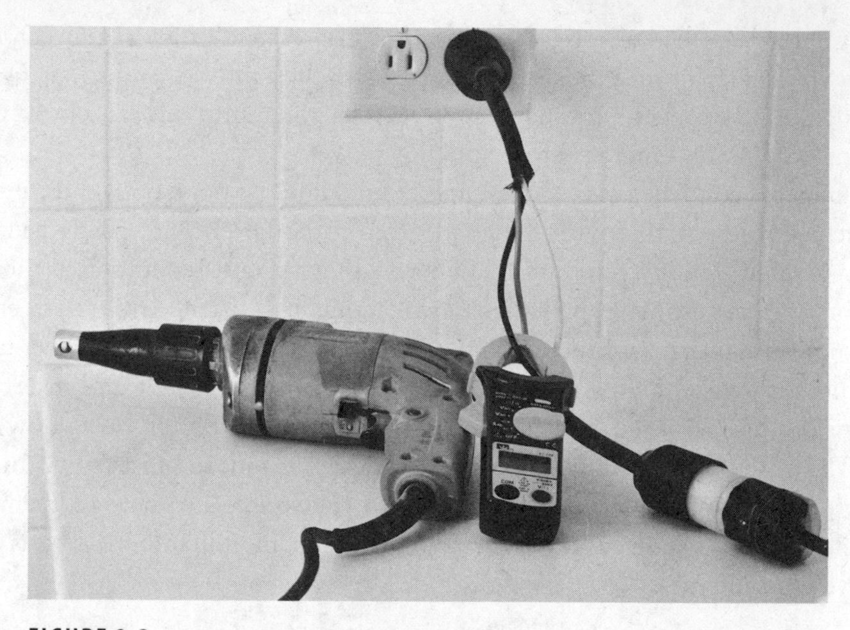

FIGURE 9-2 • Current drawn by a portable tool can be measured using a clamp-on ammeter with a fabricated splitter.

and cut away the slit jacket. This gadget provides access to the single conductor needed for clamp-on ammeter measurements.

The old analog models work fine, but a newer digital meter is perhaps a little easier to use, and it is more compact, so it can be used in tight places. Some digital models have additional features such as a hold button that will cause the meter to lock onto the highest reading over any period of time. The meter can be left in place overnight or for a weekend to reveal abnormal usage. If you have three digital clamp-on meters with hold features, a three-phase service or motor can be monitored over a period of time to detect maximum imbalance.

Still Struggling

A clamp-on ammeter determines current flow by measuring the magnetic field around a conductor. This measurement has no significant effect on the circuit.

There are other types of meters that are useful for quick measurements when the precise voltage level is not an issue. A quick, inexpensive, and easy-to-use voltage indicator is the neon test light. It will not measure precise voltage amounts, but you can get an idea of whether there are 120 or 240 V and whether it is alternating current (ac) or dc. It is good for looking over an entrance panel to see whether bus bars and individual breakers are live and to troubleshoot refrigeration, motors, relays, and controllers when you want to ascertain the presence or absence of voltage.

The neon test light consists of a small, very high-impedance neon bulb inside a molded-plastic case that also contains a resister in series with the bulb, further limiting the current. You can easily identify neutral and hot conductors by touching one probe to the wire or terminal in question and the other probe to any conductive object at some ground potential. Before and after working on electrical equipment, you can check the housing to see whether it is energized due to an internal fault. The neon test light never burns out and is overall a very useful device.

TIP ***Some electricians test for voltage by touching one neon test light probe to the object in question and touching the other probe to their finger. Because they are grounded, the bulb will light if there is voltage. This is not usually hazardous because the neon bulb draws a minute amount of current, but if there Is a short, you could get a shock, so the procedure is not recommended.***

You can make a simple voltage indicator by screwing an appliance bulb into a plug-in lamp base. It is excellent for checking receptacles to see whether they are live. A multimeter or neon test light is not as good for receptacle testing because you never know for sure when the probes are making contact. Several of these bulb indicators can be left in receptacles in a large room and viewed from a distance, useful for troubleshooting or labeling circuit breakers.

Many electricians like the solenoid voltage tester, made by various manufacturers but commonly known by its original trade name, *Wiggy*. Operating from line power, it requires no battery. It displays common voltage levels but not precise amounts. What is good about this instrument is that if you need to keep your eyes on the probes and circuit under test, you can tell when you get voltage by a distinct buzzing sound or a single loud click for dc. In a noisy environment, you can feel the Wiggy vibrating, which can be quite useful.

Some models incorporate continuity testing capability, made possible by an internal battery. They are useful for testing fuses in live circuits because switching is not necessary to go from one mode to the other. Because current is required to operate the solenoid, the Wiggy is a low-impedance meter, and it will load high-impedance circuits. This property may be put to use for testing ground-fault circuit interrupters (GFCIs). When you touch the probes to the hot terminal and the equipment ground or some other grounded conductive surface (not the neutral), the GFCI will trip.

The Wiggie should be connected to a live source only momentarily. If left in circuit continuously, it will overheat. The instrument is very compact and rugged, built to last indefinitely. While no substitute for the more versatile multimeter, it is good for specific applications and deserves a place in your toolbox.

Still Struggling

There are many types of multimeters, each useful in unique situations. Without good measuring tools, you are working in the dark.

The receptacle tester, sometimes called a *circuit analyzer*, sells for under $10 and is an essential instrument for verifying that receptacles, once energized, have been correctly wired and are free of arc faults. With a rugged plastic case and three prongs that plug into a receptacle, the instrument has three neon lights. Makes vary as to details, but the basic idea is if the three bulbs are glowing in the correct configuration, for example, only the two outer ones lighted, the receptacle is correctly wired. Any other configuration indicates a fault. There is a key printed on the plastic body permitting the user to interpret the results. If none of the neon bulbs glows, there is no power to the receptacle. Other faults include missing equipment ground, neutral and equipment ground reversed, hot and neutral reversed, and so on. If the bulbs glow weakly or flicker, there may be a loose connection or arc fault. Rapping lightly on the wall or wiggling other devices may bring out intermittents, which should be investigated and repaired.

It is a good idea to test all receptacles in a new installation and when evaluating old work, as in a home inspection. Use of this instrument will show the home or business owner that you are conscientious about your work.

Loop Impedance Meter

The loop impedance meter is a surprisingly versatile instrument that makes possible a much more comprehensive investigation into installed electrical work. European countries under the jurisdiction of the International Electrotechnical Commission (IEC) require a loop impedance meter test on all new work. Such a requirement has been slow coming to the United States, but it seems to be on the horizon, and electricians have the option to perform these tests on a voluntary basis.

Recent catastrophic weather events including flooding have exposed installed wiring to moisture. After conductors and equipment dry, corrosion sets in. Electrical connections acquire increased impedance and generate I^2R heat, giving rise to more impedance, corrosion, and heat. At a certain level, combustible material can ignite, accompanied by the possibility of loss of property or life.

Besides corrosion, other problems can become part of the equation. Terminations that are not fully torqued or are torqued to the point of failure may introduce a series arc fault. GFCIs, while they are valuable lifesaving devices, will fail to detect a break in equipment-grounding continuity, setting the stage for shock hazard.

There are a great many other faults that can be detected by the loop impedance meter. Some manufacturers have not wanted to introduce a U.S. model because voltage and frequency differences would require new design work. Megger has come out with a universal model that works everywhere, and Ideal's SureTest is a very useful instrument that has been around for years.

Loop impedance meters measure characteristics of installed electrical work and display the results so that if faults are discovered, corrective actions may be taken. Various models differ widely. What they have in common is that they are line-powered, and they introduce a load to simulate a real operating environment. The tests are of brief duration, less than a single ac cycle, so premises loads are not affected.

These definitions provide some insight into how these meters work.

Loop. An electrical path from phase to neutral and ground, often dividing into unintended segments. Unlike a circuit, which conforms to the original design, a loop may include harmful parallel ground paths. Any ground-loop impedance opposes overcurrent-device function.

Impedance. Total opposition to current flow, including resistance and capacitive or inductive reactance. These latter elements vary with frequency so that square waves and arc faults, with very fast rise and fall times, may set the stage for unwanted harmonics, voltage spikes, and surprising impedance fluctuations.

PROBLEM 9-2

What is the difference between a loop and a circuit?

SOLUTION

Generally speaking, a circuit is intentional, whereas a loop is unintentional.

One way to begin testing is to connect the instrument to the last receptacle in the branch circuit. If a fault is reported, work back upstream until the fault goes away, and that will pinpoint the location. Next, connect at the panel to ascertain available fault current. These are just two of the numerous procedures that are easy to perform yet very informative.

The Ideal SureTest 61-165, which Grainger sells for $330, though not called a loop impedance tester by the manufacturer, offers an advanced array of features, including

- Arc-fault testing
- Tests for shared neutrals
- Measures voltage drop with 12-, 15-, and 20-A loads
- Measures true root-mean-square (RMS) voltage
- Measures line voltage
- Measures peak voltage
- Measures frequency
- Measures ground-to-neutral voltage
- Measures ground impedance
- Measures hot and neutral conductor impedances
- Verifies proper wiring in three-wire receptacles
- Identifies false grounds
- Tests GFCIs for proper operation
- Conducts testing without disturbing sensitive loads
- Verifies isolated grounds (with the 61-176 adapter)

The instrument is less than 6.5 in. tall, 3 in. wide, and less than 1.5 in. deep. It comes with a two-year warranty.

Megger has solved the compatibility problem by bringing out the LT300 series high-current-loop impedance testers. Because these models operate at 16 to 400 Hz at various voltages, they work on U.S. and Canadian wiring as well as that found in Europe. At $625 to $850 depending on the exact model, this instrument performs IEC-mandated tests and is useful for U.S. electricians and inspectors. Features include

- Overcurrent devices are not tripped by testing.
- Three-wire low-current nontripping loop impedance tests provide measurements from 0.01 Ω to 2 kΩ with a resolution of 0.01 to 10 Ω.

- Two-wire high-current test is provided where GFCI connection is not an issue.
- All tests are autoranging with no range changing needed.
- Three-phase safe. The instrument is not damaged if connected across phases.
- The LT310 operates over a voltage range of 100 to 280 V for single-phase testing.
- The LT320/330 operates over a voltage range of 50 to 480 V for single- and three-phase testing.
- Frequency and phase rotation can be measured with the LT320/330. Supply voltage is measured.
- Prospective fault current is displayed.
- Light-emitting diodes (LEDs) indicate correct supply and test lead connections.
- Meters are rubber-armored for durability.
- A user's guide in the lid provides basic information.

If certification is required, spreadsheets can be imported into Microsoft Excel. For this purpose, there is a USB connection. *The LT330 can save over 1,000 test results to internal memory.*

Using an Oscilloscope

An oscilloscope differs from a multimeter in that it displays the waveform graphically rather than merely providing a numerical measurement. As with many graphic representations, there is an x and a y axis intersecting (usually) at the center of the screen. Divisions along the horizontal x axis correspond to time, and divisions along the vertical y axis correspond to volts.

A probe and ground clip permit the technician to connect to various terminals, wires, or traces on a printed circuit board and to sample the electrical energy so that its waveform is displayed on the screen. (Old models had cathode-ray tubes with vertical and horizontal deflection plates to move the beam of electrons, but of course, today's scopes have flat screens.) For the electrical power available at a premises receptacle or entrance panel, you will see the beautiful image of a sine wave. At the terminals of a motor, it may be distorted due to the nature of the load, and this will provide a wealth of information about load, source, and power transmission line.

If the screen were very wide, you could watch the waves progress across the screen, left to right, indefinitely until they ran out of space at the edge of the display. Because of the limited screen size, however, it is necessary to retrace to the left side and begin anew, all waveforms superimposed and appearing as a single static image. The problem, however, is that most of the time the waves would not merge, and the display would appear as a meaningless blur of light. This difficulty was solved by early researchers by means of *triggering*, whereby the waveforms are synchronized in response to the peak voltage or some other characteristic of the signal that is chosen.

Triggering and, a little later, the development of the digital storage oscilloscope made possible the wonderful functionality of today's instrument. It has the vast electronic resources of an advanced computer and is able to capture, store, and download to your computer fleeting transients and anomalies, making it great for power quality work. Also, it is well suited for troubleshooting audio, video, and other electronic equipment. Schematic diagrams, available as free Internet downloads, often contain graphics showing waveforms that occur at various points in the circuitry when a signal is applied at the input, and by viewing them, you can figure out what is causing distortion or some other fault.

The only problem with an oscilloscope is that the model you would want, a dual-trace triggering digital storage machine with high-frequency response and a full complement of probes, likely will cost around $5,000. If you find this sort of equipment appealing, you may want to find employment with a high-end research or manufacturing organization that is well capitalized.

Besides measuring instruments, there are other specialized electricians' tools, some essential and some optional, good to have on the job but not absolutely required. Of course, you will need the normal carpenters' and mechanics' tools. You should have everything you need to measure, cut, and fasten pieces of plywood and framing lumber. Rather than nailing, it is preferable to screw wooden pieces in place. A cordless drill with screwdriver bits is one of the frequently used tools of the trade.

Part of the job description includes repairing electrical equipment, ranging from small battery-powered items to large industrial machinery that may weigh several tons. For this, you'll want wrench and socket sets with ¼- through ¾-in. drives, plus Torx, Allen, and adjustable wrenches and Vise-Grips in a range of sizes. To start, you do not need everything. Larger, more expensive tools can be acquired as needed for specific jobs. Later, you will want a tap and die set, a laser level, an oxyacetylene torch, and similar more advanced equipment.

At the outset, build a good pair of sawhorses. They should be small, low to the ground, but very sturdy so that you can place planks across them for a working platform. Cut plywood gussets to brace the ends, and make shelves under the main cross-members. Fasten conduit hangers to the legs so that a segment of EMT can make one of the sawhorses serve as a wire caddy. A long V-block, sawed out of 2 × 2 lumber and fastened to a sawhorse, will make cutting conduit much easier.

As for specific electricians' tools, your electrical distributor will give you catalogs, or you can look online. It is not recommended that you load up on cheap junk from the auction sites or online classifieds. Instead, buy top-quality tools as needed.

The most fundamental wiring is residential. To build a service, run branch circuits in Type NM (Romex) cable, and install light fixtures and appliances, only the most basic tools are needed. Other than what has been mentioned already, you will want a ½-in. electric drill with long electricians' bits, hole saws, 5- and 6-ft step ladders, an extension ladder, a trouble light, a heavy sledge hammer for driving ground rods, a hacksaw with coarse and fine blades, and so on.

To run Type NM (Romex) cable, a cable ripper is helpful for slitting the jacket without damaging conductor insulation. To build a 200-A service, the 4/0 aluminum wires are best handled using a ratcheting cable cutter and big stripper. The rounded hole at the end of a large adjustable wrench is useful for bending and shaping heavy wire.

A long-shank electricians' screwdriver is good for getting in tight places, and with the small shank, you can spin it rapidly between your fingers to wind long screws in and out rapidly.

You don't need to invest a lot in tools to get started in residential work. Commercial and industrial wiring is another story. Fortunately, if you are employed at a large facility, the company will supply the major equipment. You will be ready to work if you have a single, well-organized toolbox, copy of the current electrical code, multimeter, and clamp-on ammeter.

The principal difference between residential and commercial electrical work is that in the latter, for the most part, Type NM (Romex) cable is not used. You will work with EMT and MC, and this involves a learning curve. But because nonresidential electricians bend metal raceways and run armored cable on an almost daily basis, they quickly become adept.

In most jurisdictions, it is not permitted to work as an electrical contractor unless a master electrician license has been obtained. This involves passing a written test and accumulating several years' experience working under the supervision of an existing master electrician. It is a venerable tradition that has been in place in many trades since ancient times.

In addition to technical expertise and physical prowess, the independent electrician must possess business skills in order to survive in our variable economic climate. You have to find a business model that works for you and for your locality. The goal should be to provide a safe, reliable product that your customer can afford and that will allow you to make a fair profit. To this end, begin by deciding a base hourly rate that you will want to get for your labor. In addition, you are entitled to buy your materials at wholesale prices and mark them up when billing customer. To do this successfully, you have to buy the materials at a good price. But do not sacrifice quality. All materials, where applicable, should be listed and marked by Underwriters Laboratories (UL) or some other testing agency. If the product is listed, it may be presumed to be free of hazards and safe to use (Figure 9-3). But a problem has emerged in recent years. Some electrical products have appeared on the market with counterfeit listing agency markings. The best course of action is to avoid questionable sources—flea markets, online auction

FIGURE 9-3 • Devices for mounting wall boxes in an existing wall. All electrical supplies should be listed by a recognized testing organization.

sites, distress sales of surplus goods, and so on. Find an electrical distributor whom you can trust, visit the physical location, and discuss the issue with the owner.

In time, you should find an electrical distributor who consistently carries quality products and makes them available to you at a favorable price. You will find at least four levels of pricing. Local hardware and building-supply stores generally offer electrical supplies at a high retail price, offering the convenience that you can pick items from their inventory as needed and return them for credit if there is a change of plans. Moreover, you may be able to speak to them and get a 10 percent or better across-the-board discount and convenient credit terms.

As for price, the big-box stores such as The Home Depot usually offer electrical tools and supplies priced at midrange, often better than the retail store even with your discount. They have extensive inventories and offer sales on construction tools and equipment including testers.

The best prices, especially when you buy in sufficient quantity to make it worth their while, are obtained from electrical distributors. They have delivery trucks, and if you are on their route, it is usually possible to place a telephone order early in the morning and take delivery by noon the same day, either at your shop or at the work site.

Electrical distributors usually have knowledgeable salespeople who are eager to supply application information on new products and to help resolve difficult problems that may arise.

Still another source for obtaining electrical supplies and equipment is direct from the manufacturer. You may be able to establish yourself in one or more specialized fields such as solar photovoltaic (PV) (Figure 9-4) or submersible-pump sales, service, and installation and obtain excellent pricing, credit terms, and technical support. Many electricians specialize in certain areas, gaining experience and expertise and acquiring testing and maintenance tools so that they can excel.

TIP ***In conversations with customers, you should never mention the words* discount *and* wholesale. *In conversations with your vendor, you should always be mentioning those words.***

There are abundant opportunities for resourceful electricians. You can offer 24-hour service if you don't mind going out at night and on weekends to work on a nonutility outage, and your work will be much appreciated. Another business model is to approach commercial and industrial venues with a view to securing maintenance contracts. Some of these facilities are not large enough to warrant a full-scale electrical department, but they may have enough of an

FIGURE 9-4 • Solar PV installation has the potential for accelerating growth.

infrastructure that they would benefit from a good proactive preventive maintenance program, especially if three-phase motors are involved. Then, when it comes time to expand or add network capacity, you will be on track to pick up the work.

There are many ways to promote your business, and some of them do not require much capital outlay. If residential work is your focus, print up business cards and flyers. Post them at all convenience stores, markets, and building supplies within your work radius.

Whenever you install an entrance panel, affix a sticker with contact information.

Get in touch with real estate agents, homebuilders, Main Street merchants, bankers, and so on, offering your services. Make clear that in return they can expect to receive referrals from you. Initial contacts can be made by e-mail, direct mail, phone, or visit.

Make site signs to post at your jobs. Make signs for your truck, home, and shop. There is no limit to the ways you can promote your business. Later, you may decide to budget funds for display ads in local papers or radio spots.

Don't neglect the website. If you have the time and expertise, or if there is a teenager in the house, you should be able to establish a web presence at virtually no cost, and this will help to grow your business. Electricians don't generally lack work. It is primarily a matter of creating a good business model and developing niche specialties.

You may aspire to build a high-power multistate operation with a number of electricians in the field and with backup shop and office employees, or perhaps you will be content to work as an independent electrician, performing installations and repair in your immediate neighborhood. Between these extremes, there is a range of operating modes. Many electrical contractors have one or two helpers, a single van on the road, and a shop and office located at the primary residence.

Regardless of how you wish to proceed, you should maintain the highest ethical standards. This is true in all professions but more so for electricians, where any mistake in design or installation can introduce hazards, resulting in property loss, injury to persons, and even fatalities.

Knowledge and expertise come with experience on the job and study. The greatest single resource is the *NEC*, and undoubtedly you will reference it frequently. Besides that venerable body of mandates and informational material, there are other documents and legislated language that should be observed.

It is essential to comply with licensing and permitting regulations that have been enacted in your locality. Licensing for electricians is required in almost every jurisdiction. There are those who choose to take a shortcut or are unable to comply. Working under the table or off the books is a totally unacceptable practice in this profession, and it invites disaster for you, your workers, your customers, and the general public.

Electrical energy is a powerful entity, and while it is immensely useful and even necessary in our lives, there is also the potential for great destruction of property and injury to persons when this force is misapplied. A first necessary step in working wisely with electricity is to comply with safety standards.

Second, if you have employees, find out what requirements are in effect in your location. Do not assume that employees can be considered subcontractors in order to circumvent Workers' Compensation laws. It won't work, and in the event of a work-site injury or job-related illness, the worker will not be covered, and you will be subject to great sanctions.

The Occupational Safety and Health Administration (OSHA) has promulgated work-site regulations designed to protect workers. Fines for violations are immense, and it is the responsibility of the employer to know those requirements and observe them. It is not a defense to say, "I didn't know." Find out the requirements, and observe them. Examples are protection against arc-flash injury, especially in commercial and industrial settings, and protection from falling when working on a roof or at any height above ground.

The Environmental Protection Agency (EPA) regulates handling of materials such as refrigerant that could harm the atmosphere or environment if improperly released. Accordingly, a nationwide license is required if you open any refrigeration piping, the condenser, evaporator, or compressor. This includes adding refrigerant to a closed system. Here again, the fines for any violations are immense.

If you are involved in any sort of excavation including the installation or repair of underground electrical lines, be aware that a wetland permit must be secured in advance of doing any such work in what is known as *hydric soil*. In the United States, the Army Corps of Engineers issues guidelines for delineating wetlands, and it must be emphasized that a wetland may not exhibit surface water or indeed be wet at all, so you may want to seek advice from a county agricultural extension agency or equivalent before proceeding.

Additionally, particularly in populated areas, it is necessary before digging to verify that there are no underground utilities that could be damaged or cause injury. In many locations, you can call toll free to reach an organization (in New England, it is known as Dig Safe) that oversees such matters. This organization will notify all utilities that may have buried lines, water, gas, electrical, phone, and so on. It is then the responsibility of the utility to visit the site within a short time and mark out its lines. If this does not occur, you will not be liable for damage.

QUIZ

These questions are intended to test your comprehension of Chapter 9. The passing score is 70 percent, but try to answer them all correctly. The quiz, like most electricians' tests, is open-book, so feel free to refer to the text. Answers appear in Answers to Quizzes and *NEC* Practice Exam.

1. What circuit parameters can be measured by a multimeter?

A. Volts, ohms, and milliamps
B. Volts, ohms, and capacitance
C. Volts, ohms, and inductance
D. Volts, milliamps, and capacitance

2. What is the cause of phantom voltage?

A. Temperature change
B. Static charge
C. Proton infiltration
D. A nearby llght source

3. In the ohm function, first you should connect to a known live source.

A. True
B. False

4. The clamp-on ammeter is

A. useful for measuring very small amounts of current.
B. a fraqile instrument.
C. available in analog and digital versions.
D. mostly obsolete.

5. The neon test light measures resistance in a live circuit.

A. True
B. False

6. The receptacle tester checks premises wiring for

A. missing equipment ground.
B. neutral and equipment ground reversed.
C. hot and neutral reversed.
D. All of the above

7. The loop impedance meter

A. checks installed electrical work.
B. tests for less than a single ac cycle.
C. performs a load test.
D. All of the above

8. **Impedance is total opposition to the flow of electric current, including resistance, capacitive reactance, and inductive reactance.**
   A. True
   B. False

9. **Electricians do not**
   A. use cordless tools.
   B. check dead circuits with a voltmeter.
   C. check dead circuits with an ohmmeter.
   D. work on live circuits when it is feasible to power them down.

10. **Type NM (Romex) cable is not used in most industrial work.**
    A. True
    B. False

chapter 10

Additional Certifications: Elevators, Fire Alarm Systems, Refrigeration, and Others = Total Job Security

The electrician's trade consists primarily of designing, installing, and repairing light and power systems in buildings or outdoor structures. This is how licensing boards generally define the work. There is not necessarily a minimum voltage level. It goes more by function. There are obviously gray areas. It could be carried to the absurd if one were to be required to possess an electrician's license and submit plans for approval prior to changing a flashlight battery.

CHAPTER OBJECTIVES

In this chapter, you will

- Learn about some additional work options for electricians.
- Look at elevator installation, maintenance, and repair as a specialty for electricians.
- Examine the field of fire alarm work.
- Learn about refrigeration and programmable logic controllers (PLCs) as niche opportunities.

I asked the New Hampshire Electricians' Licensing Board if a satellite dish installer would be required to have an electrician's license, and the representative told me that the answer was no. The board had determined that a power and light issue did not exist because the modem, which powers the system, had merely to be plugged in, and the grounding process was considered benign.

Going Further

It is assumed that the reader, having come this far, has begun the licensing process. Perhaps, depending on the jurisdiction, you have secured an apprentice card and begun amassing the work hours required prior to taking the journeyman's exam. Others will already possess that license and be on track to "mastering out." Whatever your status, there is no reason that you should not be gainfully employed and acquiring new knowledge and expertise every day.

In this chapter, we shall look at the subject from a slightly different angle. Experienced master electricians generally do not lack work, but with the chronic downturns and economic dislocations that have been occurring worldwide, it is possible to be dissatisfied with the quantity of good billable work. For a master electrician or independent contractor with employees who must provide for their families, it is disappointing to issue layoff notices. To avoid this outcome, a viable proactive expansion program could be right for you.

It is suggested that you endeavor to open new revenue streams by moving into one or more related fields. As a working electrician, you already have knowledge, expertise, and tools, so you are prepared to enter these fields immediately, albeit in an incremental fashion. In the interest of helping in the process, we shall take a look at some of these categories, and you can decide which of them will work in the context of your professional life.

Still Struggling

Many electricians hold additional licenses. Submersible-pump installation for water wells requires a license. An experienced electrician will have little difficulty qualifying and passing the exam.

Elevator Work

Elevator mechanics' licenses are required in many states, from Maine to Arizona. So the first step, if you are interested, is to do an Internet search and see what is required in your state. You may be able to secure an apprentice permit and go to work immediately for an established elevator installation and maintenance firm.

Elevator World Magazine is an excellent publication covering all aspects of the industry worldwide. Continuing education courses with easily affordable online examinations and certifications are offered.

Some schools offering elevator mechanic or related courses include

- ITT Technical Institute: www.itt-tech.edu
- DeVry University: www.devry.edu
- Kaplan University: www.kaplan.edu
- Walden University: www.waldenu.edu
- Colorado Technical University: www.coloradotech.edu

Elevator installation, maintenance, and repair are characterized by the fact that they are overwhelmingly electrical activities. If you are a working electrician and/or have knowledge, aptitude, and interest in electrical wiring and equipment, you are well placed to enter this exacting field. In *National Electrical Code* (*NEC*) Chapter 6, there is an entire article covering elevator installations from the electrical point of view.

Nearly all elevators in today's world are powered by electric motors. At one time, these motors ran on direct current (dc). Dc motors are reversible, and the speed can be controlled merely by varying the input voltage. Transitions are smooth, and dc motors perform well. The speed of an alternating-current (ac) motor, in contrast, is frequency-dependent. If you reduce the voltage, you are in effect slowing down the motor by overloading it, and this means that energy is wasted, dissipated as heat. The motor life is shortened, so this is not a viable way to control speed. (An elevator car has to slow down as it approaches each landing.)

TIP ***If you are mechanically inclined and have good computer skills, elevator work may be a valuable specialty. Expect to work long hours.***

After ac generation and distribution became dominant following the Westinghouse versus Edison confrontation, the dc motor nevertheless maintained its role in powering elevators. It could be run off ac utility power, fed through a simple rectifier.

With brushes and a commutator, however, the dc motor, despite its superb performance, remained a high-maintenance item. A big change in elevator technology came in the mid-twentieth century with the invention of the variable-frequency drive (VFD). This purely electronic solution allows the operator, human or automatic, to regulate the speed of an ac motor by varying a synthesized frequency.

Standard troubleshooting techniques work for the power supply, VFD, motor, and elevator drive (cable or hydraulic). Additionally, there is an elevator motion controller, which is an elaborate central processing unit with a user interface including an alphanumeric readout and touch keypad. A great many safety mechanisms are part of the picture, and if anything is not quite right, the motion controller responds by shutting down the elevator. It is a bit like when your desktop computer crashes and has to be restarted. If the mechanical and/or software problem persists, the controller will not allow the elevator to start, and an error message will be displayed.

An elevator technician has to be a blacksmith, a computer programmer, and everything in between. If you have a mind for this sort of thing and nerves of steel, elevator installation, maintenance, and repair might be right for you. Some electricians include elevator work as part of their job description, and they never lack work.

Fire Alarm Systems

Fire alarm systems are similar to elevators in that both have elaborate central control panels. Both these systems work well most of the time, but when there is a malfunction, it can be a real test of your expertise.

When the elevator senses an anomaly, it shuts down so that an accident or injury will not occur. When a fire alarm system malfunctions, it either false alarms or goes into the trouble state as opposed to failing to go into the alarm state when there is a real fire. For the fire alarm system, it is the job of the technician to deal with false alarms and trouble reports. With a good preventive maintenance program, they will be avoided.

The electrician is in a good position to assimilate this technology and incorporate the work into the job description.

In an urban area, like elevators, fire alarm systems are plentiful, and there is every prospect that their use will increase, especially with the possibility of sprinkler systems in every new residence. There is every reason to anticipate that the demand for fire alarm technicians will increase. Many, but not all, states license their fire alarm system designers and installers. Some states have an optional fire alarm licensing program that is not mandatory. Compliance should be seen not as a bothersome inconvenience but as a learning opportunity.

JadeLearning offers a comprehensive continuing education program for fire alarm technicians at www.jadelearning.com.

Refrigeration

Another specialty that electricians have found profitable is refrigeration. Although this type of equipment is not considered unusually hazardous, especially when compared with elevators and light and power electrical work in general, nevertheless, there are some issues that need to be considered. Of course, there is the possibility of shock hazard due to the fact that the immediate area may be damp. The original installation must include correct bonding back to the system ground of all non-current-carrying conductive surfaces that may become energized. Furthermore, as always when dealing with chemicals, we must beware of excessive exposure that may pose long-term risks.

Refrigeration technicians find that there is abundant demand for their services. Virtually all restaurants, hotels, and grocery stores have numerous refrigeration units. When one of them quits working, it is something of an emergency because there is a limited amount of time before the contents are damaged. Because in a restaurant, evening is the busiest time, refrigeration faults are likely to surface after normal working hours. By the time the technician is called and arrives on the job, it may be after midnight, so this can be a lonely profession. But there is the potential for good billable hours.

Most refrigeration repairs are electrical in nature and can be handled by an electrician, who is more likely to be available at the site. Damage to the motor-compressor can occur if it continues to run when it is overheated, there is insufficient refrigerant, or there is some sort of blockage in the refrigeration circuit. Sensors with control wiring will cause motor power to be interrupted, and the temperature in the box will rise. Electricians can deal with these problems using ordinary troubleshooting techniques. Sometimes more is required. It may be necessary to add refrigerant or open the piping to make repairs or replace components. If this is to be done, a special refrigeration technician's license is obligatory. The regulatory body is the Environmental Protection Agency (EPA). It has nationwide jurisdiction over any activity that involves risk of releasing refrigerant into the atmosphere, even if the amount is minute. This is so because worldwide these small amounts will damage the protective ozone layer in the atmosphere.

The EPA license focuses not on overall competency to install or repair refrigeration equipment but rather on handling refrigerant, adding or removing it, without releasing it into the atmosphere. The license has four levels, Types I through IV, permitting the individual to work on various sizes and types of refrigeration and air-conditioning equipment. For neglecting to comply, the fines are very high. For information on obtaining the certification, call the EPA at 1-800-886-4109, extension 229.

What Is a Programmable Logic Controller?

Any electrician who works in an industrial setting will inevitably see a programmable logic controller (PLC). Throughout the world, there are millions of PLCs. Where there is a manufacturing process that is automated, there will be a PLC.

This equipment appeared in the 1960s, first at General Motors, then in other vehicle-manufacturing plants, and soon thereafter wherever there was an assembly line of any size or complexity. Previously, vehicle makers, once a year before the new models came out, were obliged to tear apart their relay-based automation controls and start from scratch.

The PLC consists of a central processing unit with an operator interface including touchpad controls and an alphanumeric display. This nerve center is housed in a rugged enclosure, often floor-to-ceiling. The panel controls machines and industrial processes either automatically or in response to the operator's commands. The most common version consists of a modular rack-based system. It accepts input and output modules, inserted into slots by the technician. The modules are wired, often by means of Class 2 circuits, to sensors and actuators located as needed on the plant floor or even in another building.

The PLC is programmed by means of disk-based software. The technician plugs a laptop computer into the control panel and then places virtual components within an on-screen ladder diagram. In the ladder diagram, the left rail is the positive pole, and the right rail is the ground bus.

Each rung of the ladder is made up of an input and an output. Each of these entities has a numerical address. The PLC is continuously scanning the ladder, rung by rung. First, the PLC looks at the inputs, ascertaining which ones are on. Then, beginning at the top of the ladder, it executes the commands. Finally, the PLC updates the output status and begins another scan. There are two possible operating modes—programming and run.

When there is a problem with the PLC's performance, such as erratic events on the assembly line or a completely dead actuator, it is the electrician who is asked to make the initial evaluation. Large facilities will have PLC specialists. Usually whoever has the most knowledge of this equipment will be asked to troubleshoot faults or install programming as needed.

PROBLEM 10-1

How is a PLC programmed?

SOLUTION

The technician installs proprietary software into a computer (usually a laptop). A ladder diagram appears on the screen, and sensors and actuators are placed in the appropriate order. When the programming is completed, the diagram is downloaded to the control panel.

You can definitely bootstrap your way into this work. In Appendix (Books and the Internet—Using Print and Online Resources), I refer you to www.plcs.net. This site is an excellent gateway into the field, and there are numerous other Internet resources. One valuable information source is the PLC manufacturer. An example is Allen-Bradley, which is part of Rockwell Automation. Tools, and resources, a literature library, and online training are available at ab.rockwellautomation.com.

Another major player in the PLC world is Honeywell. An Internet search will take you to user forums, tutorials, and other places of interest.

This completes our survey of the electrician's trade. The goal has been to demystify some important aspects of it. The initial plan was to provide some guidance for those who intend to enter the profession or to shed light on it for those who are curious.

Electrical work is easy for motivated individuals with inquiring minds and perseverance. Figure 10-1 shows an example of the many new fields where electrical maintenance is required. There is abundant information in print and on the Internet, and it is a field in which one can learn by doing.

Electrical work is physically less demanding (at least most of the time) than other types of construction. But the stakes are very high. When you design or install electrical wiring in buildings, the lives of many people depend on your knowledge and workmanship. One bad connection or undersized conductor can initiate a fiery inferno, so everything has to be impeccable.

If there is any uncertainty regarding some facet of an installation, seek outside help. One of the best sources of information is the authority having jurisdiction (AHJ), but there is no substitute for an inquiring mind.

FIGURE 10-1 • In wind turbine installation, maintenance, and repair, the electrician's skills are in great demand.

QUIZ

These questions are intended to test your comprehension of Chapter 10. The passing score is 70 percent, but try to answer them all correctly. The quiz, like most electricians' tests, is open-book, so feel free to refer to the text. Answers appear in Answers to Quizzes and *NEC* Practice Exam.

1. Dc motors are suitable for powering elevators.

A. True
B. False

2. The speed of an ac motor for elevator application

A. does not matter because it can always be varied by gearing.
B. is frequency-dependent.
C. may be varied by means of a VFD.
D. B and C are true

3. The elevator motion controller

A. is located in a separate building.
B. incorporates many safety features.
C. is portable and may be taken to the maintenance office.
D. controls only the opening and closing of the car doors.

4. Fire alarm systems have

A. central control panels.
B. initiating devices.
C. indicating appliances.
D. All of the above

5. Refrigeration involves use of troubleshooting skills.

A. True
B. False

6. Leaking refrigerant

A. can damage the ozone layer.
B. is of concern to the EPA because it is toxic to humans.
C. is likely to cause freeze-ups in the walk-in.
D. is not important as long as it is not too costly.

7. PLCs are programmed

A. by the manufacturer.
B. only when there is a malfunction.
C. by means of a ladder diagram.
D. once every six months.

8. PLC controllers are constantly scanning the inputs.

A. True
B. False

9. An excellent source of information on PLCs is the manufacturer's documentation.

A. True
B. False

10. In electrical work, loss of property and human lives

A. is an inevitable cost of doing business.
B. can result from a single fault in the wiring.
C. is not a concern of the *NEC*.
D. is carefully investigated after a major fire.

NEC *Practice Exam*

Introduction

This practice exam is based on the 2011 *National Electrical Code* (*NEC*). As emphasized throughout this book, the current Code book is the most important item in your toolbox. Like most licensing exams, this test is open-book. While taking the test, you can refer to any of the books permitted by your board, but the *NEC* is the essential one. Either the standard edition or the *NEC Handbook* will work. You will also need a handheld calculator.

There are 80 questions. The level of difficulty is about the same as that of most licensing exams. Consult your board website to find out the rules of the game. Allow yourself the same amount of time per question. It could be three hours for 60 questions or three minutes per question.

Write the answers (just the letters) on a separate sheet, not in this book. Then, referring to the answer key in Appendix 3, score yourself. Passing score is 70 percent. The best plan would be to retake the exam as many times as necessary until you score 100 percent.

1. **The purpose of the Code is**
   A. a technical reference for designers and installers.
   B. training for novices.
   C. practical safeguarding of persons and property from hazards arising from the use of electricity.
   D. a design manual for supervisors.

2. **Compliance with the Code ensures a system that is efficient, convenient, and adequate for good service or future expansion of electrical use.**
   A. True
   B. False

3. **Items covered by the Code include**
   A. utility lines and transformers.
   B. yards, lots, carnivals, and industrial substations.
   C. installation in ships, railway rolling stock, and automotive vehicles.
   D. All of the above

4. **The Code arrangement consists of**
   A. an introduction and 12 chapters.
   B. nine chapters, each covering a different class of occupancy.
   C. annexes with specific mandates.
   D. general mandates in Chapters 1 through 4, specific occupancies in Chapters 5 through 7, information on communications systems in Chapter 8, and Tables in Chapter 9, plus nonmandatory annexes.

5. **Which of the following statements is true?**
   A. Code requirements may be waived by the authority having jurisdiction.
   B. Once enacted, a Code requirement may never be changed.
   C. Permissive rules must be followed on all job sites.
   D. Informational notes are to be rigidly enforced.

6. **The number of circuits in an enclosure**
   A. helps to prevent short circuits and ground faults.
   B. is not limited by the Code.
   C. is limited to 20.
   D. applies only to service equipment.

7. **In the Code, metric and inch-pound units**
   A. are both shown.
   B. are not relevant because trade dimensions are always used.
   C. are always exactly equivalent.
   D. All of the above

8. **Soft and hard conversions**
   A. may be used interchangeably in all cases.
   B. have no relation to safety.
   C. apply to trade sizes.
   D. None of the above

9. **Extracted material from other sources**
   A. is superseded by the Code.
   B. may be edited in the Code to conform to Code style.
   C. is not relevant to electrical design and installation.
   D. must always be found in the original documents.

10. **Mandatory Code rules identify specifically required or prohibited actions.**
    A. True
    B. False

11. **Equipment that is readily accessible includes**
    A. that which may be reached by means of a ladder.
    B. equipment above a suspended ceiling.
    C. an entrance panel in an equipment room.
    D. Any of the above

12. **Ampacity is defined as the maximum current in amperes that a conductor can carry continuously under the conditions of use without exceeding its temperature rating.**
    A. True
    B. False

13. **Which of the following statements is true?**
    A. Continuous duty is operation at a substantially constant load for an indefinitely long time.
    B. Intermittent duty is operation for alternate periods of load, no load, and rest.
    C. Periodic duty is intermittent operation in which the load conditions are regularly recurrent.
    D. All of the above.

14. **The temperature rating associated with the ampacity of a conductor shall be selected and coordinated so as not to exceed the lowest temperature rating of any connected termination, conductor, or device.**
    A. True
    B. False

15. **Which of the following statements is true?**
    A. Termination provisions of equipment for circuits rated 100 A or less or marked for 14 through 1-American Wire Gauge (AWG) conductors are to be used for conductors rated 60°C (140°F).
    B. Termination provisions of equipment for circuits rated over 100 A or marked for conductors larger than 1 AWG are to be used for conductors rated 75°C (167°F).
    C. Conductors with higher temperature ratings may be used provided that the equipment is listed and identified for use with such conductors.
    D. All of the above.

16. **On a four-wire, delta-connected system where the midpoint of one phase winding is grounded, only the conductor or bus bar having the higher phase voltage to ground is to be marked**
    A. green.
    B. red.
    C. orange.
    D. black.

17. **Arc-flash hazard warning labels are required in nondwellings for**
    A. switchboards.
    B. panelboards.
    C. meter socket enclosures.
    D. All of the above

18. **Working space and dedicated equipment space are identical.**
    A. True
    B. False

19. **Dedicated equipment space extends to a height of ______ above the top of the equipment or to the structural ceiling, whichever is lower.**
    A. 3 ft
    B. 6 ft
    C. 6.5 ft
    D. 8 ft

20. A suspended ceiling is not considered a structural ceiling when calculating dedicated equipment space.

A. True

B. False

21. Size 6 AWG or smaller insulated grounded conductors are to be identified by

A. a continuous white outer finish.

B. three continuous white stripes along the conductor's entire length on other than green insulation.

C. a continuous gray outer finish.

D. Any of the above

22. Size 4 AWG or larger insulated grounded conductors are to be identified by

A. a continuous white outer finish

B. a continuous gray outer finish.

C. three continuous white stripes along the conductor's entire length on other than green insulation.

D. Any of the above

23. Which of the following statements is true?

A. Main bonding jumpers are required in all enclosures.

B. Ungrounded systems require equipment-grounding conductors.

C. Two ground rods are required at all services.

D. Where a building is supplied by a feeder, no grounding electrode is required.

24. Vehicle-mounted generators always require a ground rod.

A. True

B. False

25. High-impedance grounded neutral systems

A. are used in many upscale residences.

B. are used where line-to-neutral loads are not served.

C. require a bare conductor from the neutral point of the generator or transformer to the grounding impedance.

D. require a grounded system conductor that must be 4 AWG copper or larger.

26. Underground metal water pipe
   A. is no longer permitted as a grounding electrode.
   B. must be in contact with the earth for 5 ft to qualify as a grounding electrode.
   C. must be in contact with the earth for 10 ft to qualify as a grounding electrode.
   D. must be copper to qualify as a grounding electrode.

27. Concrete-encased electrodes
   A. must be 20 ft long.
   B. if made up of multiple segments, must be bonded by means of copper jumpers.
   C. if bare copper conductors, must be not smaller than 6 AWG.
   D. must be encased in 4 in. of concrete.

28. Where a supplemental grounding electrode is a rod, pipe, or plate electrode, that portion of the bonding jumper that is sole connection to the grounding electrode is not required to be larger than 6 AWG copper.
   A. True
   B. False

29. Plate electrodes must be at least ____ in. below the surface of the earth.
   A. 12
   B. 24
   C. 30
   D. 36

30. Where used outside, aluminum grounding electrode conductors are not to be terminated within ____ in. of the earth.
   A. 16
   B. 18
   C. 20
   D. 24

31. Conductors of the same circuit, including the grounded conductor and all equipment-grounding and bonding conductors, must be run in the same raceway, cable, trench, or cord except
   A. all parallel installations.
   B. parallel installations underground in nonmetallic raceways.
   C. appliance cords.
   D. transformer secondaries.

32. Which conductors can occupy the same raceway or cable?

A. AC and DC circuits 600 V or less.

B. Conductors over 600 V.

C. Power-limited fire alarm and power and lighting circuits.

D. Service conductors and communication wiring.

33. In both exposed and concealed locations, where a cable- or raceway-type wiring method is installed through bored holes in joists, rafters, or wood members, holes are to be bored so that the edge of the hole is not less than ____ in. from the nearest edge of the wood member.

A. 1

B. 1¼

C. 1½

D. 1¾

34. Where raceways contain ____ AWG or larger insulated circuit conductors, and these conductors enter a cabinet, box, enclosure, or raceway, the conductors are to be protected by an insulated fitting.

A. 4

B. 6

C. 8

D. 10

35. The interiors of enclosures or raceways installed underground are to be considered wet locations.

A. True

B. False

36. Underground minimum cover required for rigid conduit, except in certain specified locations, is

A. 2 in.

B. 4 in.

C. 6 in.

D. 12 in.

37. Underground service conductors that are not encased in concrete must have their locations identified by a warning ribbon at least ____ in. above the underground installation.

A. 6

B. 8

C. 10

D. 12

38. Conduits or raceways through which moisture may contact live parts are to be sealed at ____ end(s).

A. both
B. either
C. neither
D. A or B

39. Raceways are to be provided with expansion fittings where necessary to compensate for thermal expansion and contraction

A. under driveways only.
B. for services only.
C. not for communications circuits.
D. in all applications.

40. Cable wiring methods are not to be used as a means of support for other cables, raceways, or nonelectrical equipment.

A. True
B. False

41. A single-phase motor with an inverse time breaker for branch-circuit short-circuit and ground-fault protection is to have the protective device setting at ____ percent of full-load current.

A. 100
B. 125
C. 200
D. 250

42. A motor control circuit must be capable of carrying the main power current.

A. True
B. False

43. For torque motors, the rated current is

A. full-load current.
B. locked-rotor current.
C. derived from horsepower tables.
D. not applicable to conductor sizing.

44. Motor markings are to include

A. the manufacturer.

B. rated volts.

C. full-load current.

D. All of the above plus other information

45. Locked-rotor indicating code letter H means that the kilovolt-amperes per horsepower with locked rotor is

A. 4.5–4.99.

B. 5.0–5.59.

C. 5.6–6.29.

D. 6.3–7.09.

46. Motors are to be located so that adequate ventilation is provided. An exception is

A. an elevator motor.

B. a submersible-pump motor.

C. a clock motor.

D. a stepper motor.

47. Conductors that supply a single motor used in a continuous-duty application are to have an ampacity of not less than ____ percent of the motor full-load current rating.

A. 80

B. 100

C. 125

D. 150

48. For a multispeed motor, the selection of branch-circuit conductors on the line side of the controller is to be based on the highest of the full-load current ratings shown on the motor nameplate.

A. True

B. False

C. Neither of the above

D. A or B

49. Conductors for small motors are not to be smaller than ____ AWG.

A. 16

B. 14

C. 12

D. None of these

50. Motor overload protection is not required where power loss would cause a hazard, as in the case of fire pumps.

A. True

B. False

51. Associated nonincendive field wiring apparatus may be

A. an electrical apparatus that has an alternative type of protection for use in the appropriate hazardous (classified) location.

B. an electrical apparatus not so protected that is not to be used in a hazardous (classified) location.

C. Either of the above

D. Neither of the above

52. A combustible dust is any finely divided solid material that is ____ microns or smaller in diameter and presents a fire or explosion hazard when dispersed and ignited in air.

A. 120

B. 220

C. 320

D. 420

53. A control drawing

A. is provided by the manufacturer.

B. is provided by the architect.

C. does not show interconnections.

D. is required for all hazardous areas.

54. Explosion-proof equipment is enclosed in a case that is capable of withstanding an explosion of a specified gas or vapor that may occur within it and of preventing the ignition of a specified gas or vapor surrounding the enclosure by sparks, flashes, or explosion of the gas or vapor within and that operates at such an external temperature that a surrounding flammable atmosphere will not be ignited thereby.

A. True

B. False

55. Hermetically sealed equipment is sealed against the entrance of an external atmosphere by means of

A. soldering.

B. brazing.

C. welding.

D. Any of these

56. Acetylene is a member of Group

A. A.
B. B.
C. C.
D. D.

57. A nonincendive circuit is a protection technique not permitted for

A. Class I, Division 1.
B. Class I, Division 2.
C. Class II, Division 2.
D. Class III, Division 1.

58. For grounding and bonding in Class I, Divisions 1 and 2 locations, locknut-bushing and double-locknut types of contacts are not permitted.

A. True
B. False

59. In Class II locations, ventilating pipes for motors, generators, or other rotating electrical machinery or for enclosures for electrical equipment are to

A. lead directly to a source of clean air outside buildings.
B. be screened at the outer ends to prevent the entrance of small animals or birds.
C. be protected against physical damage and against rusting or other corrosive influences.
D. All of the above

60. Enhanced bonding techniques are required in Class ____ locations.

A. I
B. II
C. III
D. All of the above

61. Branch circuits that supply neon tubing installations are not to be rated in excess of ____ amperes

A. 20
B. 30
C. 40
D. 50

62. **Where the opening of a crane control circuit would create a hazard, such as, for example, the control circuit of a hot-metal crane, the control-circuit conductors are to be considered as being properly protected by the branch-circuit overcurrent devices.**
    A. True
    B. False

63. **The feeder demand factor for four elevators is**
    A. 0.95.
    B. 0.90.
    C. 0.85.
    D. 0.82.

64. **Conductors and optical fibers located in elevator hoistways may be installed in**
    A. rigid-metal conduit.
    B. electrical metallic tubing.
    C. Type MC cable.
    D. Any of the above

65. **In an elevator machine room, required lighting**
    A. may be GFCI protected.
    B. must be GFCI protected.
    C. must not be GFCI protected.
    D. may be on the motor control circuit.

66. **A dedicated circuit is to supply ____ on each elevator car.**
    A. heating
    B. air conditioning
    C. Neither of these
    D. Both of these

67. **At least one 125-V, single-phase, 15- or 20-A duplex receptacle is to be provided in each**
    A. machine room.
    B. control room.
    C. car.
    D. A and B but not C

68. A nonmotor generator arc welder with a duty cycle of 70 has what multiplication factor?

A. 0.89
B. 0.84
C. 0.78
D. 0.71

69. In swimming pool installations, all 15- and 20-A, single-phase, 125-V receptacles located within ___ feet of the inside walls of the pool are to be protected by a ground-fault circuit interrupter.

A. 10
B. 15
C. 20
D. 25

70. In swimming pool areas, a copper conductor grid, installed to reduce voltage gradients, must be ____ AWG minimum.

A. 4
B. 6
C. 8
D. 10

71. Branch-circuit overcurrent devices in emergency circuits are to be accessible to authorized persons only

A. True
B. False

72. A Class 2 circuit

A. considers safety from a fire-initiation standpoint.
B. provides acceptable protection from electric shock.
C. Both of these
D. Neither of these

73. A Class 1 power-limited circuit is to be supplied from a source that has a rated output of not more than ____ volts.

A. 12
B. 20
C. 30
D. 600

74. Class 1 circuit overcurrent protection for 16-AWG conductors may not exceed ____ amperes.

A. 7
B. 10
C. 15
D. 20

75. Class 1 circuits and power-supply circuits are permitted to occupy the same cable, raceway, or enclosure

A. never.
B. always.
C. where the circuits are functionally associated.
D. where the circuits have the same overcurrent protection.

76. In enclosures, Class 2 and 3 circuits are permitted to be installed in a raceway to separate them from Class 1 non-power-limited fire alarm and medium-power network-powered broadband communication circuits.

A. True
B. False

77. In elevator hoistways, Class 2 and 3 circuit conductors are to be installed in

A. rigid metal conduit.
B. rigid nonmetallic conduit.
C. electrical metallic tubing.
D. Any of these

78. Cable substitutions are permitted in Article 725 where a less restricted cable is substituted for a more restricted cable.

A. True
B. False

79. Which of the following requires special listing?

A. Plenum signaling raceways
B. Riser signaling raceways
C. General-purpose signaling raceways
D. All of the above

80. CL3X is

A. Class 3 limited-use cable.
B. Class 3 power-limited tray cable.
C. Class 3 riser cable.
D. Class 3 plenum cable.

Answers to Quizzes and NEC *Practice Exam*

Chapter Quizzes Answer Key

Chapter 1
1. B
2. D
3. A
4. D
5. B
6. C
7. A
8. D
9. A
10. A

Chapter 2
1. A
2. B
3. C
4. D
5. B
6. A
7. B
8. D
9. A
10. D

Chapter 3
1. B
2. C
3. A
4. C
5. B
6. D
7. B
8. C
9. A
10. B

Chapter 4
1. B
2. B
3. C
4. A
5. B
6. D
7. B
8. C
9. B
10. D

Chapter 5
1. A
2. A
3. D
4. C
5. B
6. B
7. C
8. C
9. A
10. C

Chapter 6
1. D
2. B
3. B
4. B
5. C
6. D
7. B
8. C
9. D
10. B

Chapter 7
1. A
2. A
3. D
4. A
5. B
6. B
7. B

8. A
9. D
10. B

Chapter 8

1. D
2. A
3. C
4. B
5. D
6. C
7. D
8. B
9. D
10. D

Chapter 9

1. A
2. B
3. B
4. C
5. B
6. D
7. D
8. A
9. D
10. A

NEC Practice Exam Answer Key

1. **C.** This statement has appeared in many editions of the Code and has great bearing on what is and is not included in it.
2. **B.** The Code is concerned with electrical safety issues, not efficiency or adequacy for future needs.
3. **B.** Substations are covered if they are customer (not utility) owned.
4. **D.** The Annexes do not include enforceable mandates.
5. **A.** The authority having jurisdiction (AHJ) is frequently the local electrical inspector.
6. **A.** Overcrowded enclosures cause electrical shock and fire hazards.
7. **A.** The trend in recent years has been a move toward more use of metric measurements.
8. **D.** Most electricians' licensing exams deal in inch-pound units.
9. **B.** A complete knowledge of outside sources such as the *National Electrical Safety Code* is not necessary to pass an electricians' licensing exam, but it is a good idea to become somewhat familiar with them.
10. **A.** In addition to mandatory rules, there are permissive rules and informational notes.
11. **C.** *Readily accessible* and *accessible* are terms that recur throughout the Code.
12. **A.** Ampacities of conductors in various installations are given in Article 310.
13. **D.** Continuous duty equates to more heat in conductors.
14. **A.** Terminations will be hot spots in a circuit unless sized correctly.
15. **D.** These provisions are frequent subjects for exam questions.
16. **C.** This type of system has higher voltage to ground on the phase opposite to the winding that is grounded at its midpoint.
17. **D.** At electrical installations, inspectors check for compliance with this requirement.
18. **B.** These terms have separate meanings and different requirements.
19. **B.** This requirement protects the equipment from damage.
20. **A.** Dedicated equipment space must extend above the suspended ceiling.

21. **D.** Older Code editions referred to natural gray.
22. **A.** Ampacities of conductors in various installations are given in Article 310.
23. **B.** Equipment grounding conductors facilitate overcurrent protection.
24. **B.** A ground rod is optional.
25. **B.** These provisions are frequent subjects for exam questions.
26. **C.** When PVC waterpipe has been substituted, the grounding is not effective.
27. **A.** At electrical installations, inspectors cannot always check for compliance with this requirement.
28. **A.** A larger bonding jumper is not required.
29. **B.** This requirement reduces ground resistance.
30. **B.** Most grounding electrode conductors are copper.
31. **B.** Running hot and neutral wires in close proximity help to minimize inductive heating because the two conductors are 180 degrees out of phase.
32. **A.** All conductors must be insulated for the highest voltage present in the raceway or cable.
33. **B.** The purpose of this requirement is to protect the wiring from screws and nails.
34. **A.** An exception states that the insulated fitting is not required where threaded hubs or bosses provide a smoothly rounded or flared entry for conductors.
35. **A.** Wet-location conductors such as THWN must be used in underground conduit.
36. **C.** Rigid conduit may be used where bedrock is encountered to comply with minimum cover requirements.
37. **D.** Service conductors are not protected against short circuits and ground faults by breakers in the entrance panel.
38. **D.** Water entering an entrance panel will cause severe damage.
39. **D.** Inspectors check installations for compliance with this requirement.
40. **A.** This is a frequent Code violation.
41. **D.** This may seem strange until you realize that motors may have reduced short-circuit and ground-fault protection because they have separate overload protection. This arrangement allows them to start without interrupting the circuit.
42. **B.** It carries the electric signals directing the performance of the controller.
43. **B.** A torque motor can operate at 0 rpm when required.
44. **D.** Rated full-load speed, rated temperature rise, and time rating are among parameters appearing on the nameplate. They are considered in motor wiring design.
45. **D.** Motors with code letters nearer the beginning of the alphabet have higher locked-rotor impedance.

46. **B.** Excess heat causes damage to motor insulation.
47. **C.** The motor could be rated for continuous duty, but it could be used in some other application.
48. **A.** Multispeed ac motors usually have separate windings for the different speeds.
49. **B.** Several exceptions permit smaller sizes under limited conditions.
50. **A.** Short-circuit and ground-fault protection is nevertheless required.
51. **C.** In this apparatus, the circuits are not necessarily nonincendive themselves, but they affect the energy in nonincendive field wiring circuits and are relied on to maintain nonincendive energy levels.
52. **D.** Combustible dust is or may be present in a Class II area.
53. **A.** This may be a drawing or document.
54. **A.** Explosion-proof equipment is not appropriate for Class II locations.
55. **D.** Hermetically sealed equipment may be nonincendive.
56. **A.** It is necessary that equipment be identified not only for class but also for the specific group of the gas or vapor that will be present.
57. **A.** Class I, Division 1 is a highly restrictive location.
58. **A.** Bonding jumpers with proper fittings or other approved means of bonding are to be used.
59. **D.** Any obstruction or misapplication in ventilating pipes will lead to overheating.
60. **D.** Ordinary bonding techniques could result in dangerous arcing or loss of ground continuity.
61. **B.** Neon tubing may be filled with various inert gases.
62. **A.** If a hazard would not be created, control circuits must be protected against overcurrent.
63. **C.** This value decreases to 0.72 for 10 or more elevators.
64. **D.** A hoistway is a sensitive location and requires specialized wiring methods and materials.
65. **C.** This requirement is to minimize the possibility of loss of lighting.
66. **D.** There have been instances of passengers trapped in elevator cars for extended periods of time.
67. **D.** Electrical power is needed for maintenance and servicing in these areas.
68. **B.** The multiplication factor that times the rated primary current determines the ampacity of the supply conductors.
69. **C.** GFCIs have been great lifesavers as their use has become more widespread.
70. **C.** The wire must be bare solid copper.
71. **A.** Reliability is a large factor in emergency systems.

72. **C.** Class 2 circuits are less hazardous than Class 1 and Class 3 circuits, and the wiring methods are more permissive.
73. **C.** Class 1, 2, and 3 circuits are defined by the capacity of their power sources and by their usage.
74. **B.** Smaller conductor sizes are permitted here.
75. **C.** Signaling circuits require protection from external power sources.
76. **A.** The raceway provides necessary separation.
77. **D.** Liquid-tight flexible nonmetallic conduit and intermediate metal conduit are also permitted.
78. **A.** The issue is fire propagation and smoke generation.
79. **D.** Nonmetallic raceways could themselves become sources of smoke generation and fire propagation.
80. **A.** X denotes limited use. This cable is less expensive but not suitable for many applications.

Appendix: Books and the Internet—Using Print and Online Resources

I have stated that the most important item in an electrician's toolbox is the *National Electrical Code* (*NEC*), revised every three years and published by National Fire Protection Association (NFPA). The book, in its various formats, is available at a discount from Amazon (see the link at www.electriciansparadise.com) or directly from the publisher. The *NEC Handbook* contains the complete text of the current Code plus lengthy commentary, many photographs, and diagrams. The *NEC* is the foremost print resource, and you will refer to it on a regular basis.

I would be remiss if I did not draw attention to my two previous McGraw-Hill books. They are *2011 National Electrical Code Chapter-by-Chapter* and *Troubleshooting and Repairing Commercial and Electrical Equipment*.

Stan Gibilisco is a noted and very prolific technical author with a vivid writing style and the ability to convey concepts with great clarity and precision. He has written about 50 books on a variety of technical topics, all published by McGraw-Hill. Most of his works focus on mathematics and electrical theory. They do not assume prior technical knowledge, yet they take you far into the topics. If you need some review, these titles are highly recommended:

- *Teach Yourself Electricity and Electronics*
- *Mastering Technical Mathematics* (with Norman Crowhurst)
- *Algebra Know-It-All: Beginner to Advanced and Everything in Between*

- *Physics Demystified*
- *Electricity Demystified*

Jack Benfield's *Benfield Conduit Bending Manual*, 2nd ed., published by EC&M Books, is a concise yet definitive guide to the art and science of metal raceway bending.

The Art of Electronics, by Paul Horowitz and Winfield Hill, is a highly successful, wonderful treatment of digital and analog circuit design. It is not overly mathematical, instead emphasizing the intuitive and creative aspects of the subject. A hefty 1,152 pages, it is published by Cambridge University Press. A new edition should be out late in 2013.

There are some excellent periodicals of interest to electricians, some free to qualifying electrical contractors:

- *Electrical Construction and Maintenance Magazine* (*EC&M*) has been an industry standard for over a century. Heavily *NEC*-oriented, it runs articles on wiring methods, power quality, new technology, and current business trends.
- *Electrical Business Magazine*, a Canadian periodical, provides coverage of industry news as well as how-to articles for electrical installers and designers. Back issues are archived at www.ebmag.com.
- *Elevator World* provides excellent worldwide coverage of leading-edge elevator technology. It includes an excellent continuing education program with lots of technical information that can lead to certification in this demanding field.
- *Electrical Connection Magazine*, an Australian publication, contains how-to articles on all phases of electrical work. Heavy on new technology, this lively publication provides an interesting perspective from that vibrant part of the world.
- *Electrical Contractor Magazine* covers all types of power, voice, data, and video work. Published by the National Electrical Contractors Association, this periodical is strong on Code, new technology, and analysis of industry trends.

When making use of the Internet, you have to exercise judgment. There is a lot of erroneous material out there, and much of it is superficial, with feckless discussions of whether a receptacle ground terminal should be at the top or bottom. Nevertheless, the Internet contains a huge amount of good information, and nowhere is this more true than in the field of electrical and electronics work.

Here are some of the most useful sites:

- JadeLearning.com offers online licensing exam preparation classes for electrical professionals. It also provides continuing education for fire alarm technicians and Code change courses for electricians. These courses are approved by many state licensing boards and meet requirements for electricians to maintain their licenses.
- Electriciansparadise.com, my site, archives articles on Code, electrical construction techniques, and related topics.
- Plcs.net offers a course on DVDs—all about programmable logic controller (PLC) programming and maintenance. You can sign up for a free *PLC Tips Newsletter* delivered to your e-mail inbox every two to three weeks. PLCs are described and discussed in Chapter 10.
- Mikeholt.com is a gigantic site with lots of free information for electricians. Mike Holt is the world's foremost electrical educator, and he maintains an impressive web presence where you can sign up to receive an excellent free newsletter.
- Rfcafe.com is an amazing and highly informative website with lots of electronic information, much of it relevant to working electricians.
- Ieee.org is the website of the Institute of Electrical and Electronic Engineers (IEEE). This organization is the world's largest professional association for the advancement of technology. There are many benefits for members. Dues are $185 per year, $32 for students.
- UL.com is the website for Underwriters Laboratories, which tests electrical and other supplies and equipment and certifies them from the point of view of safety. For an interesting overview, visit the site and click on "History."
- Hamuniverse.com has lots of information on shortwave radio, much of it of interest to electricians.
- NFPA.org is the website of the National Fire Protection Association, the authority on fire, electrical, and building safety. This organization issues a new edition of the *National Electrical Code* every three years. It also publishes a number of other codes and standards.
- BenfieldDirect.com is the definitive source for information on bending conduit. Available on this site are DVD courses, the *Benfield Conduit Bending Manual*, and Ideal conduit bending tools.

- Repairfaq.org, Silicon Sam's enormous site, is full of information and links to the world of electronics and electrical construction. You will find articles on everything from microchips to lasers.
- Electronics-lab.com contains many archived technical articles, links, and online forums.
- Solorb.com/elect is a large site featuring electronics circuits, a used-equipment garage sale, and many links to related sites.
- Circuit-fantasia.com presents an unbelievable existential approach to electronics.
- Eserviceinfo.com contains a large amount of service information on electrical equipment you may be called on to repair. Schematics can be downloaded, so if you have a printer, here is your chance to build a library.
- Eehot.com is an advanced electronics forum with threaded comments and the ability to upload images and files, archived equations, and a lot more.
- EDX.org is the website of a joint venture of Massachusetts Institute of Technology (MIT) and Harvard University. It offers free online courses, many on electronics and electrical engineering. Additional colleges and universities are expected to join this open-syllabus trend in the near future so that people all over the world will have access to a wealth of knowledge.
- Thefoa.org is the website of the Fiber Optic Association (FOA), a non-profit organization that offers much information, news of the industry, and a guide to what the FOA has available to help you in your work. (See Chapter 7 for more information on fiber optics as an addition to the electrician's trade.)

I hope these links are helpful. This merely scratches the outer surface of the Internet, a body of knowledge that is growing every day. In many ways, it is the true gateway into the electrician's trade.

Index

A

B

D

E

F

G

H

I

N

O

P

R

V

W

Z